Grewell/Benatar/Park
Plastics and Composites Welding Handbook

Plastics and Composites Welding Handbook

Edited by
David A. Grewell, Avraham Benatar, Joon B. Park

With contributions by
C. Bonten, C. Brown, F. Chipperfield, J. P. Dixon, I. Froment, T. Hutton,
E. Pecha, P. Rooney, A. Savitski, C. Tüchert, R. Wise, C.-Y. Wu

HANSER
Hanser Publishers, Munich
Hanser Gardener Publications, Inc., Cincinnati

The Editors:
David A. Grewell, The Ohio State University, Columbus, OH 43221-3560
Avraham Benatar, The Ohio State University, Columbus, OH 43221-3560
Joon B. Park, Visteon Corporation, Dearborn, MI 48120

Distributed in the USA and in Canada by
Hanser Gardner Publications, Inc.
6915 Valley Avenue, Cincinnati, Ohio 45244-3029, USA
Fax: (513) 527-8801
Phone: (513) 527-8977 or 1-800-950-8977
Internet: http://www.hansergardner.com

Distributed in all other countries by
Carl Hanser Verlag
Postfach 86 04 20, 81631 München, Germany
Fax: +49 (89) 98 12 64
Internet: http://www.hanser.de

The use of general descriptive names, trademarks, etc., in this publication, even if the former are not especially identified, is not to be taken as a sign that such names, as understood by the Trade Marks and Merchandise Marks Act, may accordingly be used freely by anyone.

While the advice and information in this book are believed to be true and accurate at the date of going to press, neither the authors nor the editors nor the publisher can accept any legal responsibility for any errors or omissions that may be made. The publisher makes no warranty, express or implied, with respect to the material contained herein.

Library of Congress Cataloging-in-Publication Data

Plastics and composites welding handbook / edited by David A. Grewell, Avraham Benatar, Joon B. Park.
    p. cm.
ISBN 1-56990-313-1 ("hardcover" : alk. paper)
1. Plastics–Welding–Handbooks, manuals, etc. I. Grewell, David A. II. Benatar, Avraham. III. Park, Joon B.
    TP1160.P53 2003
    620.1'923–dc21

                        2003003986

Bibliografische Information Der Deutschen Bibliothek
Die Deutsche Bibliothek verzeichnet diese Publikation in der Deutschen Nationalbibliografie;
detaillierte bibliographische Daten sind im Internet über <http://dnb.ddb.de> abrufbar.
ISBN 3-466-19534-3

© Carl Hanser Verlag, Munich 2003
Production Management: Oswald Immel
Cover illustration by Richard Teynor
Coverdesign: MCP • Susanne Kraus GbR, Holzkirchen, Germany
Typeset, printed and bound by Kösel, Kempten, Germany

Contributors List

Avraham Benatar
The Ohio State University
Columbus, OH 43221-3560

Christian Bonten
Universität Essen
45141 Essen
Germany

Chris Brown
TWI Limited
Cambridge CB1 6AL
United Kingdom

Felicity Chipperfield
TWI Limited
Cambridge CB1 6AL
United Kingdom

James P. Dixon
Alloyd Co. Inc.
DeKalb, IL 60115

Ian Froment
TWI Limited
Cambridge CB1 6AL
United Kingdom

David A. Grewell
The Ohio State University
Columbus, OH 43221-3560

Ted Hutton
ATOFINA
Wetmore, CO 81253

Joon B. Park
Visteon Corp.
Dearborn, MI 48120

Ernst Pecha
Leuze GmbH & Co.
Germany

Paul Rooney
Branson Joining Technologies
Honeoye Falls, NY 14472-9798

Alexander Savitski
Edison Welding Institute
Columbus, OH 43221

Carsten Tüchert
Universität Essen
45141 Essen
Germany

Roger Wise
TWI Limited
Cambridge CB1 6AL
United Kingdom

Chung-Yuan Wu
Visteon Corporation
Dearborn, MI 48120

Preface

Welding methods are an important class or group of joining technologies for manufacturing of plastic and polymeric composite products. As the use of plastics and composites increases in a variety of applications so do the needs for rapid, reliable, and high-quality welding methods. Also, as polymers and polymeric composites are being used in more structurally demanding applications, the structural requirements for the welds increase as well. Since welding is usually done near the end of the manufacturing cycle, there is significant value added to the products before these final joining operations. Therefore, it is important to carefully consider and select the appropriate welding technique for the application. This requires a fundamental understanding of the physical and phenomenological aspects of the welding steps that are common to all welding processes as well as understanding the fundamentals of each welding method. In addition, one must consider process selection strategies and methods for evaluation of the properties of the welds. With this information, it is then possible to consider and select suitable welding methods during product design and manufacturing stages.

This handbook is intended for a wide audience including: welders, welding equipment operators, design engineers, manufacturing engineers, chemical engineers, material scientists, and research and development personnel. Each chapter was developed by experts in the field with a wide breadth of information dealing with all the welding aspects including materials, process phenomenology, equipment, and joint design. The authors also included many application examples to assist engineers in considering the materials and geometries that were successfully used for a specific process. While not all encompassing, the handbook provides most of the basic information along with references that can be used to gather additional information or details.

We are indebted to the many people that were involved in the preparation of this handbook. First, we would like to thank the many authors that shared their expertise in their respective chapters. We thank the numerous companies that shared welding applications and photographs of parts and equipment. Special thanks go to Ed Samuels and Bonnie Stephens from Branson Ultrasonics Corp., Dagmar Ziegler from Wegener North America, Inc., and Steve Chookazian from Ashland Specialty Chemical Company Emabond Systems for being extremely helpful. We are especially grateful to Dr. Christine Strohm, our editor from Hanser Publishers, for encouraging and prodding us throughout this project. We also thank Christine for the many hours, sometimes during late nights and weekends, that she spent editing and improving our writing so that it was more readable and understandable.

Columbus, OH	David Grewell
Columbus, OH	Avraham Benatar
Detroit, MI	Joon Park

Contents

1   Introduction

Avraham Benatar

1.1   Joining of Plastics and Composites

The use of plastics and composites is increasing in a variety of applications including auto-motive, aerospace, appliance, infrastructure, packaging, electronics, toys, and more. Plastics and composites offer many advantages including:

- High specific strength (strength/density)
- High specific modulus (modulus/density)
- Design flexibility
- Reduced manufacturing costs
- Corrosion resistance
- Excellent life expectancy

The wide variety of polymers and polymeric composites makes it possible to select and at times even customize the material to the application. Therefore, these materials are being used in more and more applications under more stringent requirements. Continuing efforts in discovering new polymers while modifying and reinforcing existing polymers to create new material combinations further increases the potential uses for these materials.

Joining is a critical step in the manufacture of components from plastics and polymeric composites. As the requirements for the component increase, so do the requirements for joining, especially in structural applications. Joints are needed when:

- Part integration is impossible due to complexity and/or cost,
- Using different materials in one structure,
- Disassembly is required, and
- Repair of damage is needed.

Careful consideration must be given to the joint or the benefits gained from using specific materials may be eroded at the joint. For example, if lower weight or corrosion resistance were an important consideration in material selection, it would be a determent to the final assembly to use metallic fasteners for joining. There are many important considerations for joint design in plastics and composites including:

- Region and magnitude of load transfer across the joint,
- Type of loading applied to the joint,
- Geometry of the parts,
- Operating environment,
- Reliability and repeatability,
- Anticipated life.

The methods for joining plastics and composites can be divided into three major categories: mechanical joining, adhesive bonding, and welding (see Fig. 1.1). Mechanical joining involves the use of separate fasteners such as metallic or polymeric screws or it relies on integrated design elements that are molded into the parts such as snap-fit or press-fit joints. In adhesive bonding, an adhesive is placed between the parts (adherents) where it serves as the material that joins the parts and transmits the load through the joint. In welding or fusion bonding, heat is used to melt or soften the polymer at the interface to enable polymer intermolecular diffusion across the interface and chain entanglements to give the joint strength. Each of these categories is comprised of a variety of joining methods that can be used in a wide range of applications. This handbook is devoted to welding processes only, as there are similar handbooks devoted to adhesive bonding and mechanical joining.

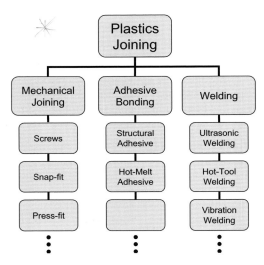

***Figure 1.1** Plastic and composites joining processes*

1.2   **Plastics and Composites**

Plastics are materials that are derived from synthetic polymers. "Polymer" comes from the combination of the Greek words "poly", which means many, and "méros", which means parts. Therefore, a polymer is a very long molecule (macromolecule) that is made-up of many repeating parts that are usually referred to as "mers". Similarly, an oligomer is a molecule made up of a small number of mers (typically ten). For example, consider a macromolecule of polyethylene. As shown in Fig. 1.2, ethylene gas molecules are the monomers that are linked together to form oligomeric ethylene and polyethylene molecules.

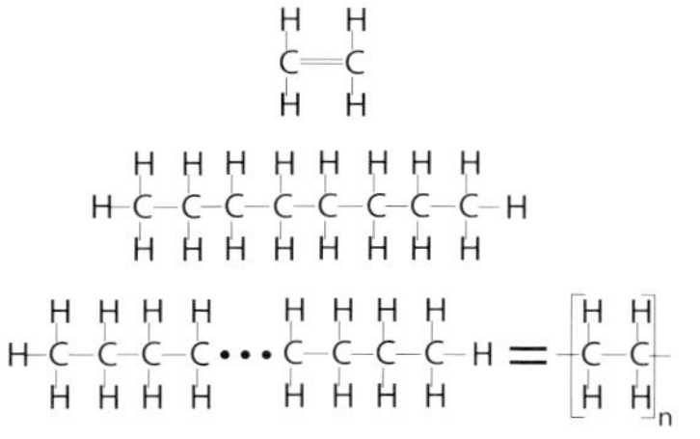

Figure 1.2 *Ethylene monomer (top), oligomer (middle), and polymer (bottom)*

It is possible to create a variety of polymers by changing the mer group. Tables 1.1 and 1.2 show examples of vinyl and vinylidene compounds. It is also possible to have molecules with different bonds between the carbon atoms on the backbone of the chain (like double and triple bonds) as well as other atoms or chemical groups instead of some of the carbons. Copolymers have more than one type of mer group in the molecules; for example ABS molecules have acrylonitrile, butadiene, and styrene mer groups linked together to form a long molecule. Therefore, the number and types of polymers are almost endless.

Polymers are divided into two major groups: thermosets and thermoplastics (see Fig. 1.3). As the names imply, thermosets are thermally set to their final geometry while thermoplastics can be thermally heated to enable reforming or reshaping. A long polymer molecule, which is also referred to as chain, has mer groups linked together by primary chemical bonds like the carbon-carbon bond for polyethylene (see Fig. 1.2).

Thermoplastics are made up of many polymer chains that are held together by weak secondary chemical bonds. Therefore, when a thermoplastic is heated, the weak secondary bonds between molecules are broken and the molecules can slide relative to each other allowing the polymer to be reshaped.

For a thermoset, the long polymer chains are held together by primary chemical bonds called cross-links (see Fig. 1.4). Therefore, after the chemical reaction that forms the cross-links is completed, reheating cannot break these bonds without also breaking the carbon-carbon bonds with the chain; at that point, chains break into smaller segments resulting in degradation. Thermosets can also be created by branching (see Fig. 1.4). For welding or fusion bonding, it is desirable to enable molecules to diffuse across the weld interface without degradation. Therefore, welding or fusion bonding of thermosets is not possible except in rare instances where cross-links can be broken without degradation.

Table 1.1 Examples of Vinyl Compounds

Mer Group	R
Ethylene	H
Vinyl Chloride	Cl
Propylene	CH_3
Acrylonitrile	C N
Styrene	

Table 1.2 Examples of Vinylidene Compounds

Mer Group	R'	R''
Isobutylene	CH_3	CH_3
Vinylidene Chloride	Cl	Cl
Methylmethacrylate	CH_3	$COOCH_3$

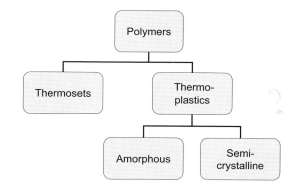

Figure 1.3 *Types of polymers*

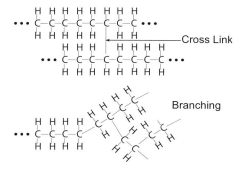

Figure 1.4 *Thermosets are formed by cross-linking or branching*

Thermoplastic molecules are held together by secondary chemical bonds, which can be broken by heating. This permits thermoplastic molecules to slide relative to each other and to diffuse across a weld interface. Therefore, thermoplastics can be welded. As shown in Fig. 1.3, thermoplastics can also be divided into two groups: amorphous and semi-crystalline. Amorphous thermoplastics have long molecules that are coiled, kinked, or have large side groups so that they are randomly intertwined without any order or structure. Semi-crystalline thermoplastics have long linear molecules forming highly ordered and compact structures in the form of crystallites. However, these very long molecules cannot form highly ordered structures throughout, resulting in some amorphous regions in the material, hence the name semi-crystalline.

Amorphous polymers behave like glassy solids below a temperature known as the glass transition temperature (T_g). Above the glass transition temperature, an amorphous polymer softens, permitting reshaping. While semi-crystalline polymers also have a glass transition temperature, they also have a melting temperature (T_m), denoting the point where the crystallites melt. Therefore, many semi-crystalline polymers can be used in structural applications at temperatures above the glass transition temperature, so long as the temperature is

kept below the melting temperature. For example, for HDPE the glass transition temperature is at about –120 °C, but it provides useful solid properties to temperatures in excess of
75 °C. Figure 1.5 shows that the viscoelastic modulus of an amorphous polymer drops
rapidly above T_g. For a semi-crystalline, the viscoelastic modulus drops slightly above T_g,
with a much larger drop above T_m.

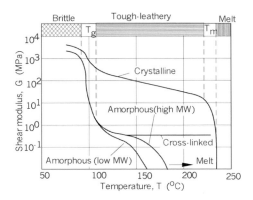

Figure 1.5 *Shear modulus for amorphous, semi-crystalline, and cross-linked polymers*

Composites are materials that consist of two or more distinct materials or parts. These may
include polymers with one or more types of fibers and particulates. Organic and inorganic
fibers can be in the form of continuous fibers, long fibers, or short fibers and they can have
specific alignment or orientation or they can have random orientation. The particulates can
also be organic or inorganic in the form of spheres, plates, ellipsoids, hollow and solid
particles. Therefore, the combination of a large variety of polymers with many types of
fibers or particulates can result in an almost endless list of composites. The chemical,
mechanical, and optical properties of all polymers can be changed greatly through additives, which in addition to reinforcements may also include plasticizers, flame retardants,
or colorants. These changes in chemical, physical, and mechanical properties can have an
important impact on joining aspects.

1.3   Classification of Welding Processes

Welding processes are often categorized and identified by the heating method that is used.
All processes can be divided into two general categories: internal heating and external
heating (see Fig. 1.6). Internal heating methods are further divided into two categories:
internal mechanical heating and internal electromagnetic heating.

External heating methods rely on convection and/or conduction to heat the weld surface. These processes include: hot tool, hot gas, extrusion, implant induction, and implant resistance welding.

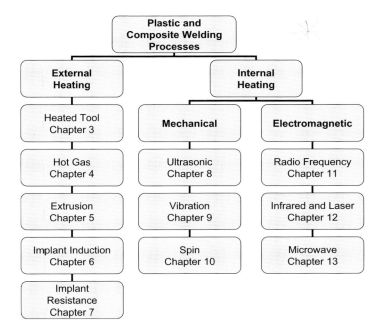

Figure 1.6 Classification of plastic and composites welding processes

Internal mechanical heating methods rely on the conversion of mechanical energy into heat through surface friction and intermolecular friction. These processes include: ultrasonic, vibration, and spin welding.

Internal electromagnetic heating methods rely on the absorption and conversion of electromagnetic radiation into heat. These processes include: infrared, laser, radio frequency, and microwave welding.

1.4 Goals of the Handbook

This handbook was developed to provide the user with a resource for information about all welding methods. Each chapter was developed by experts in the field with a wide breadth of information dealing with all the welding aspects including materials, process phenomenology, equipment, and joint design. The authors also included many application examples to assist engineers in considering the materials and geometries that were successfully used for a specific process. While not all encompassing, the handbook provides most of the basic information along with a long list of references that can be used to gather additional information or details.

1.5   How to Use the Handbook

The handbook is divided into six major sections:

1. Fundamentals of plastics welding – Chapter 2

2. Welding methods that rely on external heating (see Fig. 1.6)
   a. Heated Tool (Hot Plate) Welding – Chapter 3
   b. Hot Gas Welding – Chapter 4
   c. Extrusion Welding – Chapter 5
   d. Implant Induction (Electromagnetic) Welding – Chapter 6
   e. Implant Resistance Welding – Chapter 7

3. Welding methods that rely on mechanical internal heating
   a. Ultrasonic Welding – Chapter 8
   b. Linear and Orbital Vibration Welding – Chapter 9
   c. Spin Welding – Chapter 10

4. Methods that rely on electromagnetic internal heating
   a. Radio Frequency Welding – Chapter 11
   b. Microwave Welding – Chapter 12
   c. Infrared and Laser Welding – Chapter 13

5. Process Selection – Chapter 14

6. Testing of Welds – Chapter 15

While careful review of the whole book is highly encouraged, reading of a few select chapters is also possible as each chapter was prepared to stand on its own. All readers would benefit from reading Chapters 1, 2, 14, and 15. To learn more about a specific topic, the reader should use the references at the end of each chapter. In some cases, a list of suggested reading is also provided at the end of the chapter.

1.6   Suggested Reading on Polymers

Van Vlack, L.H., *Elements of Materials Science and Engineering* (1975) Addison-Wesley, Chapter 7

Fried, J.R., *Polymer Science and Technology* (1995) Prentice Hall, New Jersey

Billmeyer, F.W, *Textbook of Polymer Science* (1984) John Wiley & Sons, New York

1.7   Suggested Reading on Composites

Agarwal, B.D. and Broutman, L.J., *Analysis and Performance of Fiber Composites* (1990) John Wiley & Sons, New York

Delmonte, J., *Technology of Carbon and Graphite Fiber Composites* (1981) Van Nostrand Reinhold, New York

Schwartz, M.M., *Composite Materials Handbook* (1984) McGraw-Hill, New York

Lubin, G., *Handbook of Composites* (1982) Van Nostrand Reinhold, New York

2 Fundamental Steps in Plastics and Composites Welding

Avraham Benatar

2.1 Introduction

Welding of thermoplastics and thermoplastic composites is the process of joining parts through heating in order to melt (or soften) and fuse the polymer at the interface (or faying surface). Sometimes this process is also known as fusion bonding because the strength buildup at the interface is due to intermolecular diffusion of the long polymer molecules across the interface. While this description is simplistic, it captures the essence of thermoplastic welding processes. If one examines welding processes, it is possible to identify five distinct steps that make up these processes. For some welding processes, these steps are sequential while for other processes they may occur simultaneously. For example, consider the five steps involved in a process like hot plate welding, which is described on more detail in Chapter 3. First, the surfaces may be cleaned or machined to insure that they are free of contaminants and that they match the heated plate and the part to be welded; this can be regarded as a surface preparation step. Then there is a heating step, where the parts are brought in contact with the hot plate to heat and melt the polymer on the surface of both parts. Next, a pressing step is initiated, where the hot plate is retracted, and the molten surfaces on the parts are pressed together. Once sufficient contact is achieved between molten polymer layers from the two parts, the fourth step of intermolecular diffusion occurs. In this case, at the same time the fifth step of cooling of the molten polymer at the weld interface also begins. With this summary of the welding process reviewed, each of the steps is detailed in the following sections.

2.2 Surface Preparation

Surface preparation is the first step in getting thermoplastic or thermoplastic composite parts ready for welding. Surface preparation is especially important when using manual or semi-automated processes because the level of manual handling required in these processes increases the likelihood of contamination. Typically, surface preparation involves machining and/or cleaning. Machining of the surface is used to insure that it is square, or that it fits the weld surface from the other part, or in some cases it may be required to apply a groove or chamfer like in hot gas welding. Typically, cleaning is done using a solvent to degrease the surface. However, in extreme cases more extensive mechanical and/or chemical cleaning may be used. For example, surface abrasion (mechanical) followed by solvent wash may be used. In other cases, chemical etching to remove the contaminated surface layer followed by solvent wash may be used. However, when surface texture or surface properties need to be preserved, mechanical or chemical surface treatments are not possible.

For automated welding processes, surface preparation is rarely used. In those cases, the major source of contamination is the mold release agent transferred to the part during molding. Therefore, in these cases, and in general, the best solution is to establish a release layer on the mold that does not transfer to the part; this is done by applying a thermoset release layer to the mold surface and allowing it to cure and adhere onto the mold surface. Benatar and Gutowski [1] evaluated the effect of mold release on the ultrasonic welding of thermoplastic composites. They found that even for non-transferable mold release, some transfer did occur to the surface of the part. However, if standard curing procedures for the mold release were followed, the amount of transfer was small and it did not have a noticeable effect on ultrasonic welding of the parts. When mold release was cured directly on the surface of the thermoplastic composites, it was impossible to ultrasonically weld them. More studies are required in this area to determine the effect of different types of surface contaminations on welding.

2.3   Heating

There are several ways to heat thermoplastics and thermoplastic composites for welding. With all welding processes, only the surface near the weld interface is heated in order to melt or soften the polymer because:

- It is more energy efficient to heat and melt a thin layer than a large amount of material.
- It is faster to melt the polymer near the joint area than to wait for convection or conduction to melt the polymer everywhere since plastics tend to have a relatively small thermal conductivity.
- Melting or softening the polymer everywhere would make it impossible to support the part during heating to avoid deformation or warping. For advanced composite materials with high volume fractions of fibers, heating without pressure can result in void formation within the composite in a process that is referred to as deconsolidation (see Fig. 2.1) [1]. Therefore, for these materials surface heating is even more critical, especially when the welding process permits application of pressure during heating and welding to avoid or minimize the potential for deconsolidation.

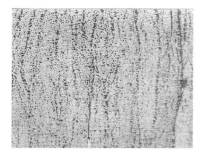

a) Consolidation by melting
with pressure

b) Deconsolidation by melting
without pressure

Figure 2.1 *Micrographs of a polymer/graphite composite after melting the polymer a) with the application of pressure b) without the application of pressure*

Welding processes are often categorized and identified by the heating method that is used. Three general categories are used:

1. External heating methods rely on convection and/or conduction to heat the weld surface. These processes include:

    a. Heated Tool (Hot Plate) Welding – Chapter 3

    b. Hot Gas Welding – Chapter 4

    c. Extrusion Welding – Chapter 5

    d. Implant Induction (Electromagnetic) Welding – Chapter 6

    e. Implant Resistance Welding – Chapter 7

2. Mechanical internal heating methods rely on the conversion of mechanical energy into heat through surface friction and intermolecular friction. These processes include:

    a. Ultrasonic Welding – Chapter 8

    b. Linear and Orbital Vibration Welding – Chapter 9

    c. Spin Welding – Chapter 10

3. Electromagnetic internal heating methods rely on the absorption and conversion of electromagnetic radiation into heat. These processes include:

    a. Radio Frequency Welding – Chapter 11

    b. Microwave Welding – Chapter 12

    c. Infrared and Laser Welding – Chapter 13

Heating is usually considered the most critical step in the welding process because welding is not possible without the formation of a thin melt or softened layer on each part. This layer is necessary to insure that flow at the interface is possible to achieve intimate contact and that intermolecular diffusion and chain entanglements are possible. As might be expected, the amount of heat input and the required temperature of the melt or softened layer differ for amorphous and semi-crystalline polymers. For amorphous polymers, it is necessary to exceed the glass transition temperature in order for flow and diffusion to occur. However, if the temperature of the softened polymer is very close to the glass transition temperature then flow and diffusion can take a long time. To keep the welding process short, the recommended heating temperatures for most amorphous thermoplastics is at about 100 °C above their glass transition temperature. For semi-crystalline polymers, it is necessary to form a melt layer exceeding the melting temperature. This usually requires higher energy input to overcome the latent heat of fusion. At temperatures below the melting temperature but above the glass transition temperature, most molecules are still bound in crystallite regions so flow and diffusion are not possible. To insure that the polymer is molten along the entire weld interface, a heating temperature exceeding the melting temperature by about 50 °C is usually recommended for semi-crystalline polymers.

Heating rates and heat transfer are important because they affect the speed of welding and the thickness of the melt or softened layer. In general, mechanical internal heating methods produce the fastest heating rates and usually have the shortest cycle time and thinnest melt or softened layers. Electromagnetic internal heating methods have intermediate heating rates, cycle times and intermediate melt layer thickness. External heating methods generally have the slowest heating rates, longest cycle times, and the thickest melt layers.

Heat transfer during welding is also important, as it will affect the rate of melting and the size of the molten or softened layer. In general, heat transfer can occur by conduction, convection, and radiation. As shown in the simplified heat transfer analysis in Fig. 2.2, thermal conduction into the part is much greater than heat transfer by convection or radiation to the surrounding air. Therefore, in most analyses of heat flow it is possible to neglect convection and radiation except when considering heating by convection or radiation. As the thermal conductivity of the fibers and particles in the composite increase, so do the heat losses from the melt surface to the interior of the part. However, by placing a thermoplastic film or a resin-rich layer at the composite surface for thermal insulation, these conduction losses can be reduced substantially. An approximate heat transfer analysis (illustrated in Fig. 2.3) shows the reduction in heat flux into graphite PEEK composite due to the insulating layer.

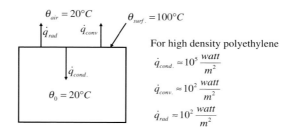

Figure 2.2 *Comparison between conduction, convection, and radiation heat losses during plastic welding*

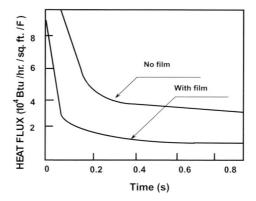

Figure 2.3 *Effect of resin rich surface (25μm of thermoplastic) on the heat flux into the composite*

Thermoplastic composites have anisotropic thermal conductivities that further complicate local heating. Figure 2.4 shows the relative thermal diffusion into a polymer, a polymer/ graphite composite with the fibers in the transverse direction, and polymer/graphite composite with the fibers in the longitudinal direction. Due to the higher thermal conductivity of the fibers, thermal diffusion into the longitudinal fiber composite is ten times greater than in the polymer alone. Figure 2.5 shows the isothermal contours developed in unidirectional graphite PEEK composite heated by a point source. For isotropic materials, these contours would be circular. However, for the anisotropic composite the contours are elliptical because the thermal conductivity in the longitudinal (fiber) direction is much greater than the thermal conductivity in the transverse direction. Therefore, when welding advanced composites it is possible that some areas that are far from the weld zone would heat to a high enough temperature to melt and possibly cause deconsolidation of the material in those regions.

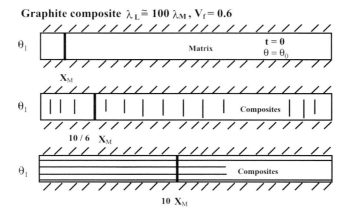

Figure 2.4 *Relative thermal diffusion distances for a polymer (top), polymer/graphite composite with fibers in the transverse (middle), and longitudinal (bottom) directions*

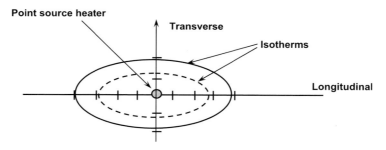

Figure 2.5 *Thermal contours in a unidirectional composite that is subjected to a point heat source*

Polymers and many polymeric composites have relatively low thermal conductivity and thermal diffusivity compared to metals or highly conductive composites such as graphite fiber composites. Therefore, conduction of heat from the melt surface into the interior of the part occurs slowly, permitting modeling of many parts as semi-infinite solids. As shown in Fig. 2.6, a semi-infinite solid extends from a surface to infinity in all other directions. In addition, for many welding processes it is possible to estimate the temperature of the surface or the heat flux at the heated surface. Chapters 3 to 13 provide more information on the phenomenology and heating that occurs with most thermoplastic welding processes. Once the surface heat flux or temperatures are determined for a given process, it is possible to use the semi-infinite solid model to estimate the temperature distribution in the part for some simple cases.

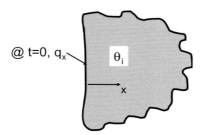

Figure 2.6 *Model for a semi-infinite solid subjected to surface heat flux*

Consider a semi-infinite solid that is subjected to a uniform and constant heat flux over the surface as shown in Figure 2.6. In this case, the temperature distribution in the solid as a function of position and time is given by the following relation [2]:

$$\theta(x,t) = \theta_i + \frac{2 \cdot \dot{q}_0}{\lambda}\left[\sqrt{\frac{\kappa \cdot t}{\pi}} \cdot \exp\left(-\frac{x^2}{4 \cdot \kappa \cdot t}\right) - \frac{x}{2} \cdot \text{erfc}\left(\frac{x}{2\sqrt{\kappa \cdot t}}\right)\right] \tag{2.1}$$

Where θ is the temperature, x is the position, t is time, θ_i is the initial temperature of the solid, $\dot{q}_0$ is the heat flux at the surface, λ is the thermal conductivity, κ is the thermal diffusivity, and erfc(z) is the complementary error function. It is important to note that the complementary error function can be calculated using most spreadsheet programs (e.g. Microsoft Excel®) or mathematical software (e.g. Mathcad® and Matlab®). Equation 2.1 can be used to estimate the temperature distribution for a number of welding processes including ultrasonic, vibration, spin, implant induction, implant resistance, infrared, laser, radio frequency, and microwave welding.

For hot gas and extrusion welding, a heated gas is used to heat and melt the surface of the parts. Therefore, in these cases convection heat transfer is important. Figure 2.7 shows a semi-infinite solid model with the surface being heated by convection using a fluid at an elevated temperature. In this case, the temperature distribution in the part is given by the following relation [3]:

$$\theta(x,t) = \theta_i + \left(\theta_f - \theta_i\right) \cdot \left[\operatorname{erfc}\left(\frac{x}{2\sqrt{\kappa \cdot t}}\right) - \right.$$

(2.2)

$$\left. \exp\left(\frac{x \cdot h}{\lambda} + \frac{\kappa \cdot t}{\left(\frac{x}{h}\right)^2}\right) K \cdot \operatorname{erfc}\left(\frac{x}{2\sqrt{\kappa \cdot t}} + \frac{\sqrt{\kappa \cdot t}}{\left(\frac{x}{h}\right)}\right) K \right]$$

Where θ_f is the fluid temperature and h is the surface heat transfer coefficient.

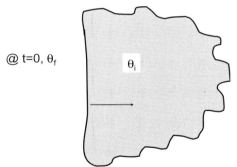

@ t=0, θ_f θ_i

Figure 2.7 *Model for a semi-infinite solid subjected to convection at surface*

For hot plate welding, the surface of the part is kept at a constant temperature during heating. In this case, the semi-infinite solid surface is kept at a constant temperature as shown in Fig. 2.8. The temperature distribution in the solid is given by the following relation [3]:

$$\theta(x,t) = \theta_i + \left(\theta_s - \theta_i\right) \cdot \operatorname{erfc}\left(\frac{x}{2\sqrt{\kappa \cdot t}}\right)$$

(2.3)

Where θ_s is the surface temperature. Notice that Equation 2.2 can be reduced to Equation 2.3 for the special case of setting the heat transfer coefficient in Equation 2.2 to infinity.

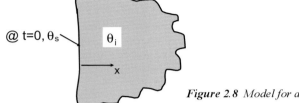

Figure 2.8 *Model for a semi-infinite solid subjected to surface temperature*

There are many advantages to estimating the temperature distribution in the parts during welding. For many processes, it is important to determine the melt layer thickness. In cases where the microstructure near the weld is important, the temperature distribution can be used to estimate the size of the heat-affected zone and the type of morphology that is likely to develop. For dimensional accuracy, the temperature distribution may be used to estimate thermal expansion and ultimate distortion in the parts. For long-term performance of the parts one might use the temperature distribution to estimate the thermal stress and residual stress levels in the welded parts.

2.4   Pressing

Once the heated zones are established, pressure must be applied to form intimate contact between the parts, while for composite parts it can also suppress deconsolidation. This is done in two phases: during the first phase surface asperities are deformed and intimate contact of the parts is achieved. During the second phase parts of the melt layer are squeezed out and any entrapped gases and contaminated polymer from the joint area are displaced. Both phases are dominated by squeezing flow mechanisms. In some cases, pressure is not directly applied during welding, rather, thermal expansion of the parts during heating results in pressure developing at the interface.

Two surfaces that are brought together will not be in intimate contact everywhere due to the surface roughness of each part (see Fig. 2.9). Even in the case of a molten polymer surface, asperity geometry will be maintained because the viscosity of the molten polymer is very high. Applying pressure across the interface will increase the contact area. For an elastic-plastic solid, increasing the pressure can cause local elastic and plastic deformation of surface asperities, thereby increasing the area of contact [4, 5, 6]. For a viscoelastic liquid like a molten polymer, increasing the pressure will also increase the contact area; however, some time-dependency is to be expected.

Figure 2.9 *Schematic of two surfaces in contact*

To consider the deformation of one asperity, it is possible to expand the faying surface as shown in Fig. 2.10. For simplicity, this model assumes that the single asperity deforms under the applied force as if it is a Newtonian liquid that is squeezed between two plates (see Fig. 2.11). To reduce the problem to a two-dimensional problem it is assumed that the parts and the droplet are very long compared to their height and width. For a Newtonian fluid with a constant viscosity that is subjected to a constant squeezing force, the following relation gives the resulting height of the droplet with respect to time:

$$\frac{h_0}{h} = \left(1 + \frac{5 \cdot F \cdot t \cdot h_0^2}{4 \cdot \eta \cdot L \cdot b_0^3} \right)^{\frac{1}{5}} \tag{2.4}$$

Where $2h_0$ is the initial height of the droplet, $2h$ is the height of the droplet at time t, F is the squeeze force, η is the viscosity, $2L$ is the length of the droplet, and $2b_0$ is the initial width of the droplet.

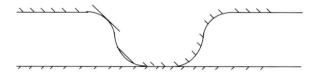

Figure 2.10 *Close-up view of a single asperity*

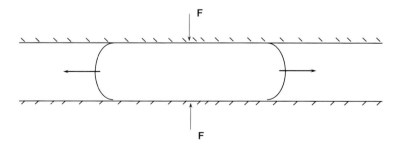

Figure 2.11 *Squeeze flow model for a single asperity*

How does this show that it is possible to expand the faying surface as shown in Figure 2.10?

A molten thermoplastic will typically exhibit the behavior of a viscoelastic material. At elevated temperatures – above the melting temperature for semicrystalline thermoplastics or above the glass transition temperature for amorphous thermoplastics – the polymer can be modeled as a power law fluid [7, 8]. Dara and Loos [7] and Gutowski et al. [8] derived the two dimensional squeeze flow equations for a power law fluid, arriving at the following:

$$\frac{h_0}{h} = \left(1 + t\frac{2n+3}{4n+2}\left(\frac{(4h_0 b_0)^{n+1} \cdot F \cdot (n+2)}{(2b_0)^{2n+3} \cdot L \cdot m} \right)^{\frac{1}{n}} \right)^{\frac{n}{2n+3}}$$

(2.5)

where m and n describe the power law relation between the shear stress τ_{xy} and the shear strain rate $\dot{\gamma}_{xy}$ as $\tau_{xy} = m \cdot \dot{\gamma}_{xy}^n$. As expected, Equation 2.5 shows the relation between the height of the droplet and the force, time, m, and (indirectly) temperature (since m usually follows an Arrhenius relation with temperature) [7, 9].

$$m = m_0 \cdot \exp\left(-\frac{E_a}{R \cdot T} \right)$$

(2.6)

where E_a is the activation energy, R is the universal gas constant, and T is the absolute temperature.

From conservation of mass, the width of the droplet is inversely proportional to the height of the droplet. In other words, as time increases the height of the droplet will decrease while the width of the droplet will increase. It is interesting to note that, to verify Equation 2.5, Dara and Loos [7] studied, theoretically and experimentally, the area of contact or the degree of intimate contact for a graphite thermoplastic composite. They found, as predicted by Equation 2.5, that the area of contact increases with increasing force and/or time. The area of contact also increases with increasing temperature through the decrease in m with increasing temperature.

It is also important to note that Equations 2.4 and 2.5 are especially well suited for ultrasonic welding; the energy director can be regarded as a single asperity which is squeezed between two rigid plates during the welding process. A similar approach may be used for other welding processes where a protrusion is squeezed between the two parts as is common in vibration welding and sometimes hot plate welding through the use of so called "weld beads".

The second phase in the pressing step is to squeeze the melt layer between the two parts as shown in Figure 2.12 for round components or Figure 2.13 for rectangular components. As part of the melt layer squeezes out, it drags with it entrapped gases (between the deforming asperities) and contaminants out of the weld region. Again, to simplify the problem it is assumed that the molten polymer can be modeled as a Newtonian liquid. Bird et al. [9] derived the relation for squeeze flow between two circular parallel disks:

$$\frac{h_0}{h} = \left(1 + \frac{16 \cdot F \cdot t}{3 \cdot \pi \cdot R^4 \cdot \eta}\right)^{\frac{1}{2}}$$ (2.7)

where R is the radius of the circular disk.

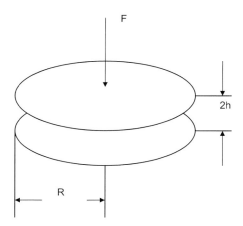

Figure 2.12 *Squeeze flow of the melt layer for round components*

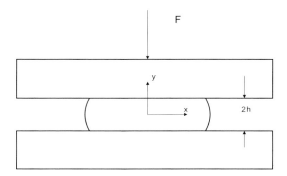

Figure 2.13 *Squeeze flow of the melt layer for rectangular components*

Similarly, the squeeze flow of the melt layer between rectangular plates can be considered. To reduce this case to a 2-D problem, one can consider a melt layer with a length that is much greater than the width and thickness (see Fig. 2.13). In this case, the following relation gives the height of the melt layer as a function of viscosity, force, and time:

$$\frac{h_0}{h} = \left(1 + \frac{F \cdot t \cdot h_0^2}{2 \cdot b^3 \cdot L \cdot \mu}\right)^{\frac{1}{2}}$$ (2.8)

While Equations 2.7 and 2.8 can actually be derived for a power law fluid model as well, assuming a Newtonian fluid is typically sufficient for most cases. In fact, the squeeze flow of polymers is a very complicated process that should include inertial effects and elastic effects. Grimm [10] provides a review of the squeeze flow of polymeric liquids. He recommends a semi-empirical approach to account for the non-Newtonian effects on the squeeze flow process. In actual welding, there are even greater complications as the temperature in the melt is not constant and in addition, it is continually changing during heating and cooling.

During welding it is desirable to have squeeze flow occur as quickly as possible. However, in some cases there are some physical limitations to the force that can be applied. And, caused by the rapid heat transfer away from the heated zones and into the parts, there also are limitations on time and temperature. In addition, high temperature may degrade the thermoplastic material. Therefore, in order to maximize the area of contact it is important to minimize the viscosity for any given force, time, or temperature. This is especially important when considering composite materials with high levels of reinforcement. Historically, various researchers [7, 8] studied the influence of fillers on viscosity. Although they do not agree as to the mechanisms for changes in viscosity, they do agree that the viscosity increases substantially for high volume fractions of fillers. Dara and Loos [7] and Gutowski et al. [8] experimentally determined that composites with fiber volume fractions of 60–70 % could have an increase in m in the power law relation of 1–3 orders of magnitude (as compared to the neat resin). During welding, such large magnitudes of viscosity can affect the degree of intimate contact and the degree of welding substantially. However, since the concern here is only with surface asperities and the melt layer, the viscosity can be decreased by making the composite surface resin rich or by placing a polymer film at the surface. This polymer film layer deforms more easily than the composites, thereby quickly achieving intimate contact between the two parts. The film layer is also advantageous for heating considerations (as discussed previously, in Section 2.3).

The above discussion is only a partial description of the problem of achieving intimate contact when bonding two thermoplastic parts. A more complete description of this process is quite complicated, due to the irregularity of the composite surface geometry and to the non-uniform temperature field that is typical for thermoplastic parts. Furthermore, the whole issue of air entrapment and squeeze-out has not yet been completely understood.

It can be expected to see dramatically different behavior between semi-crystalline and amorphous thermoplastics. Basically, semi-crystalline materials flow more easily, provided that the temperature is above the melting temperature. On the other hand, the flow behavior of amorphous polymers depends upon how high above the glass transition temperature they are heated. In general, Equations 2.4 to 2.8 provide a good start for developing a complete description for the flow of these materials; they provide good insight into the effects of time, temperature, pressure, and the power law constants for the viscosity on achieving intimate contact between two to-be-bonded parts.

2.5   Intermolecular Diffusion

Once polymer-to-polymer contact is achieved at the interface, intermolecular diffusion and entanglement is needed to complete the process and to form a good weld. Autohesion is the phenomenon describing the intermolecular diffusion and chain entanglement across a thermoplastic polymer interface forming a strong bond (see Fig. 2.14). Unlike adhesion, which relies on surface energetics (or secondary chemical bonds between (dis)similar materials), autohesion relies on chain entanglement and secondary bonds for polymer chains of similar materials. Under ideal conditions, the diffusion is complete when the interface is no longer discernible from the bulk.

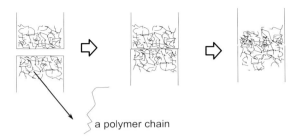

a polymer chain

Figure 2.14 Intermolecular diffusion across an interface

The autohesion process is often divided into five phases:

1. Surface rearrangement
2. Surface approach
3. Wetting
4. Diffusion, and
5. Randomization

In welding, surface rearrangement, approach and wetting can be regarded as part of the pressure step. Diffusion and randomization are part of the intermolecular diffusion step.

The motions of individual linear polymer chains are modeled using the reptation theory developed by DeGennes [11, 12]. In the reptation model, a chain is confined into an imaginary tube that represents the constraints imposed by adjacent chains. The chain is free to move in a snake like motion within the tube, but it cannot leave the tube except at the ends (see Fig. 2.15). Over a period of time, the chain can slip out of its original tube. The motion of the ends of the chains, out of the original tube, propagates toward the center of mass of the chain until the whole chain is out of the original tube and a new tube is generated. In reality, many new tubes are generated during the elapsed time period, but all of these are partially made up of the original tube (see Fig. 2.15). Therefore, there are two time scales in the reptation model: T_e is the time associated with short-range motions or wriggling of a chain within its tube, and T_r is the time required for the chain to generate a new tube. Since, for autohesion, motion outside the tube is necessary but generation of a new tube is not required, the time scale of importance for autohesion is in the range of $T_e < t < T_r$ [13]. Also, T_e is proportional to the molecular weight squared, and T_r is proportional to the molecular weight cubed [11]. Because the molecular weight is usually very large, $T_e << T_r$ and, for all practical purposes $t > T_e$ for $t > 0$. Therefore, during the time range $0 < t < T_r$ it is possible to relate the mean square path ($\langle \ell \rangle$) of the diffused chain length (ℓ) to the average interpenetration distance (X) of the chains across an interface (or even an imaginary plane in the bulk) [13].

$$X^2 \propto \langle \ell \rangle \tag{2.9}$$

During the early stages of diffusion and randomization ($t < T_r$), Kim and Wool [13] show that ℓ is proportional to $t^{\frac{1}{2}}$. Therefore,

$$X \propto t^{\frac{1}{4}} \tag{2.10}$$

Wool and O'Connor [14] proceed further to relate the interpenetration distance to the stress at fracture and time. The stress at fracture is proportional to the average interpenetration distance, therefore

$$\sigma_f \propto t^{\frac{1}{4}} \tag{2.11}$$

where σ_f is the fracture stress. They also derive relations for the time for complete healing (complete intermolecular diffusion), with respect to the molecular weight, temperature, and pressure.

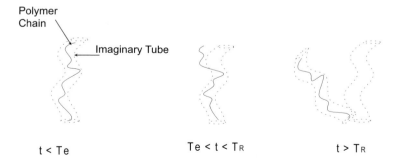

Figure 2.15 Imaginary tubes are used to describe the constraints applied onto diffusion polymer chains by adjacent chains

Many investigators [6, 13–15] have found agreement between experiments and the relations presented above. The experiments and the theoretical work presented so far have been primarily for amorphous thermoplastics. For amorphous polymers, intermolecular diffusion can take place at any temperature at or above T_g. This theory can also be applied to semi-crystalline polymers, provided that the crystals are fully melted and the polymer chains behave as random chains; this is only possible at temperatures above the melting temperature (T_m), which is usually much higher than T_g. Therefore, for semi-crystalline polymers at temperatures above T_m, intermolecular diffusion is very rapid and it is difficult to measure. Below T_m no welding or intermolecular diffusion takes place due to the crystals which bind the molecules. Therefore, in order to estimate the diffusion time for semi-crystalline polymers, the results from experiments with amorphous polymers need to be applied.

2.6   Cooling

The final step in the welding process is the cooling and re-solidification of the polymer at the joint. During this final step, semi-crystalline matrices re-crystallize to obtain their final micro-structure, while amorphous polymers retain any molecular orientation that was previously induced. In addition, thermally induced residual stresses and distortion remain frozen in the parts.

For amorphous polymers, the pressure step results in molecular orientation parallel to the weld line, which is frozen in during cooling. Figure 2.16 shows evidence of frozen-in molecular orientation in hot plate welded polystyrene by photo-elastic fringe lines near the weld interface. Triumalai and Lee [15] also showed that reheating and melting of the polymer near the weld line resulted in shrinkage transverse to the weld due to molecular orientation.

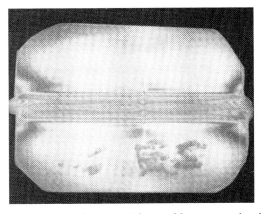

Figure 2.16 Photo-elastic fringe lines showing evidence of frozen-in molecular orientation [15]

For semi-crystalline polymers, the rate of cooling affects the rate of crystallization and the formation of spherulites in and near the weld. Figure 2.17 shows the morphology in hot plate welded polypropylene. Four distinct regions are generally observed.

- *Region 1* is the frozen skin that develops when rapid cooling of the surface occurs prior to pressing the parts together. It is generally hoped that no-frozen skin would be visible at the interface.

- *Region 2* has small spherulites caused by the rapid cooling that is experienced in this region, which does not permit the spherulites to grow in size.

- *Region 3* has deformed spherulites probably caused by partial melting and softening of the spherulites, which are then deformed during squeeze flow. Region 3 is considered the weakest as a crack can easily propagate between these aligned spherulites.

- *Region 4* shows the bulk morphology, because this region is far enough from the weld line so that it was not affected by heating and melting of the interface.

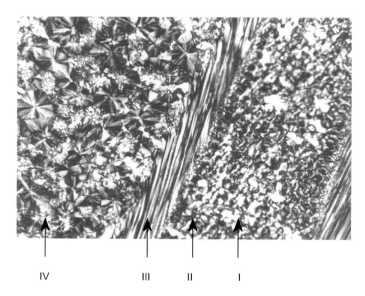

IV   III   II   I

Figure 2.17 *Morphology of hot plate welded polypropylene showing four regions: I: frozen skin, II: small spherulites, III: elongated spherulites, and IV: bulk spherulites (courtesy Edison Welding Institute)*

Distortion and residual stress formation also occur during cooling. Surface heating of the parts results in localized heating and thermal expansion, which is constrained by adjacent cool sections in the part. During cooling, molten or softened sections experience thermal contraction, which is also constrained by adjacent cooler sections in the parts. For example, in hot gas welding this results in very visible bending distortion as portrayed in Fig. 2.18. The magnitude of bending distortion in this case can be reduced by using a double-V joint where the heat input on the top and bottom is almost the same.

Formation of residual stress near the weld line is another adverse affect of welding. It can lower the weld strength, shorten fatigue life, lower fracture toughness, and result in stress corrosion cracking. Experimental and theoretical determination of the residual stress level caused by welding is difficult because of the viscoelastic nature of the polymers and the complexity of measuring residual stresses in general. Park and Benatar [16] showed that significant levels of residual stress are possible. However, additional work is required in this area to better predict and understand the effects of the process parameters on residual stress formation as well as the implication of the residual stress on the behavior of the weld.

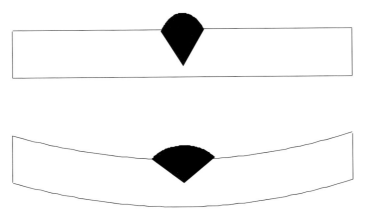

Figure 2.18 Bending distortion is typically observed during hot gas welding

2.7   References

1. Benatar, A., Gutowski, T.G., *SAMPE Journal* (1987) **Vol. 23**, No. 1, p. 33–38

2. Carslaw, H.S., Jaeger, J.C., *Conduction of Heat in Solids* (1959) Oxford University Press, London

3. Rohsenow, W.M., Choi, H., *Heat, Mass, and Momentum Transfer* (1961) Prentice-Hall, Englewood Cliffs, N.J.

4. Ling, F.F., *Surface Mechanics* (1973) John Wiley, New York

5. Suh, N.P., Turner, P.L., *Elements of the Mechanical Behavior of Solids* (1975) Scripta, Washington D.C.

6. Rabinowitz, E., *Friction and Wear of Materials* (1965) John Wiley, New York

7. Dara, P.H., Loos, A.C., "Thermoplastic Matrix Composite Model," Interim Report 57 (Sept. 1985) The NASA-Virginia Tech Composites Program

8. Gutowski, T.G., Cai, Z., Soll, W., Bonhome, L., *American Society of Composites First Conference on Composite Materials* (1986)

9. Bird, R.B., Armstrong, R.C., Hassager, O., *Dynamics of Polymeric Liquids: Volume I, Fluid Mechanics* (1987) John Wiley, New York

10. Grimm, R.J., *American Institute of Chemical Engineers' Journal* (1978) **Vol. 24**, No. 3, p. 427–439

11. DeGennes, P.G., *Journal of Chemical Physics* (1971) **Vol. 55**, p. 572

12. DeGennes, P.G., *Physics Today* (June 1983), p. 33–39

13. Kim, Y.H., Wool, R.P., *Macromolecules* (1983) **Vol. 16**, p. 1115–1120

14. Wool, R.P., O'Connor, K.M., *Journal of Applied Physics* (1981) **52** (10)

15. Triumalai, V. and Lee, L.Y. "Characterization of Polystyrenes in the Vicinity of Hot Plate Welded Joints," *Proceedings of the 48th Annual Technical Conference*, Society of Plastics Engineers, Brookfield, CT (1990)

16. Park, J. and Benatar, A. "Moiré Interferometry Measurement of Residual Strains In Implant Resistance Welding Of Polycarbonate," *Proceedings of the 50th Annual Technical Conference*, Society of Plastics Engineers, Brookfield, CT (1992)

3 Heated Tool (Hot Plate) Welding

Ernst Pecha and Alexander Savitski

3.1 Introduction

Hot plate welding is one of the most popular methods for joining thermoplastics because it is a simple, reliable, and economical way of producing strong welds. Hot plate welding works by placing the two components to be welded against or near a heated tool. The weld surface is then heated by conduction, convection, and radiation to promote melting. Once a certain amount of melt is built up at the faying surfaces, the heat source is removed and the two surfaces are brought together. Usually there is a certain amount of squeeze flow in order to assure proper fusion and help remove any dirt. The interface is then allowed to solidify resulting in a weld.

Hot plate welding is an easy process to control with wide operating windows resulting in strong joints, especially considering long-term behavior. In fact, it is common for the weld strengths to exceed the strength values of any other welding method. Often it is possible to weld unfilled materials and reach weld strengths, which approach base material strength, if parameters are correctly selected and welding machines have suitable designs. When fillers are added to thermoplastics to enhance the mechanical properties or to reduce cost, they can adversely affect welding, making it more difficult to select suitable parameters and achieve high weld strength. In these cases, more experimentation with production parts is needed to achieve the desired joint strength. Joint design is also important in order to transfer the design load through the joint.

While hot plate welding is often used in mass production, it has relatively long cycle times (10 to 30 s or even longer), which is longer than the cycle time for internal heating welding methods such as vibration or ultrasonic welding. However, hot plate welding cycle times are still comparable to injection molding cycle times, so there should be no problems for continuous production. Also, it is possible to scale-up the welding machines to simultaneously weld multiple parts to make up for longer cycle times.

Hot plate welding has been utilized since the early 1930s, when it was used for joining semi-finished products made of PVC. It gained importance with the increasing use of polyolefins, which are difficult to bond adhesively. Hot plate welding was used for pipe lines and appliances as well as for thermoplastic moldings in volume or mass production. As a consequence, nearly all-significant research institutes for plastic technology extended their studies to investigate and model hot plate welding. At the same time, various national and international associations for welding technology (e.g., the Deutscher Verband fuer

Schweissen (DVS) in Germany, the American Welding Society (AWS) in the US, and the Comité Européen de Normalisation (CEN) for European standards) compiled specifications and guidelines, which can help users set up welding equipment or troubleshoot problems.

3.2   Process Description

3.2.1   Conventional Hot Plate Welding

Hot plate welding, like other methods for sealing thermoplastic materials, provides the necessary heat and pressure to melt and fuse parts. Figure 3.1 shows the various stages of hot plate welding.

During the heating phase, the joining area is heated through conduction by direct physical contact with the hot plate. The hot plate temperature ranges from approximately 30 to

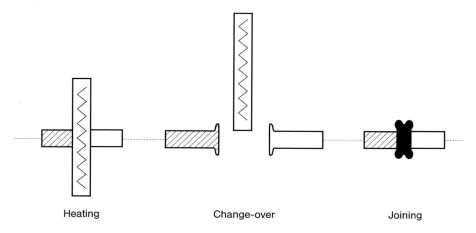

Figure 3.1 Stages of hot plate welding

100 °C above the material's melting point. In current industry practice, the joining surfaces of the parts rarely match the theoretical welding plane due to molding warpage. Therefore, during the matching phase, the weld surface is melted under high pressure (ranging between 0.2 and 0.5 MPa), forcing it to conform to the hot plate geometry, which has the desired weld geometry. The melt that is squeezed out during the matching phase is incorporated into the flash. As soon as the welding parts are in full contact with the hot plate, the pressure is reduced to a minimum (it must be assured that the welding parts and hot plate continue to remain in contact), which signifies the start of the heating time. During this phase, heat penetrates the joining area without substantial displacement of the melting material. The temperature of the melt surface continues to rise slightly and reaches approximately 20 °C below the surface temperature of the hot plate. The viscosity of the melted material as well as the amount of the plasticized layer can be precisely controlled via the temperature setting of the hot plate and the heating time. To minimize sticking of the melt to the hot plate, it is usually coated with polytetraflouroethylene (PTFE). To avoid degradation of the PTFE, the maximum hot plate temperature is limited to 250 °C.

At the end of the heating time, the change-over phase (time) starts. During this phase, the parts are removed from the hot plate, the hot plate retracts and the welding parts are joined under a predefined pressure. The change-over time should be as short as possible, since the melt cools off during this time – see Fig. 3.2.

The joining (forging) phase begins when the parts are pressed together. The joining pressure depends on melt viscosity and wall thickness of the parts. Usually the weld pressure ranges between 0.025 and 0.05 MPa. The joining pressure is maintained during the entire cooling time. The temperature drops during the cooling time due to heat flow into the surrounding air as well as into the deeper material layers. The cooling time ends as soon as the temperature of the melt has significantly fallen below the crystalline melting point (of semi-crystalline thermoplastics) or below the softening temperature (glass transition temperature of amorphous thermoplastics). Now is the time to release the pressure. The welded parts can be unlocked and removed from the machine. If the parts are cooled rapidly (e.g., by use of coolants), bad welds may occur. It is important to note that the welded moldings should not be exposed to mechanical stresses before they have completely cooled down to room temperature. The quality of a weld depends greatly on the melt layer thickness and on ensuring that some of the melt layer remains in the welding zone after joining. Applying too high a joining pressure may result in nearly complete squeeze out of the melt, which will adversely affect the weld quality.

The necessary matching and joining pressures are generated in volume production by welding machines, which are equipped with pneumatic, hydraulic or electric systems. Such machines ensure that all pressures required for the welding process are accurately controlled and reproduced. Some machines allow control of the matching and joining displacement through the use of mechanical stops rather than pressure control. Setting of these stops guarantees displacement of the melted material out of the joining area.

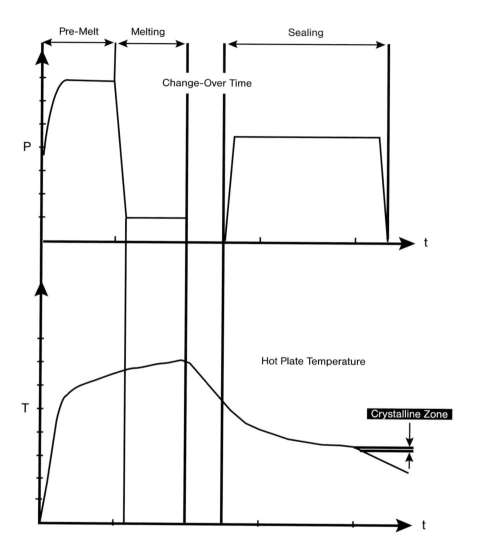

Figure 3.2 *Pressure and temperature history during hot plate welding*

3.2.2   Variants of Hot Plate Welding

As the use of conventional hot plate welding increased, variants had to be developed in order to meet special requirements of current industry practice. These variants enlarge the application field of hot plate welding and offer a number of solutions to users' problems. For example, Table 3.1 shows the different variants used to weld polyolefins. These variants of hot plate welding are discussed in more detail in the following sections.

Table 3.1 *Approximate Process Parameters for Hot Tool Welding of HDPE, LDPE and PP Products with Wall Thickness from 2 to 4mm (DVS)*

Material	Melt flow rate [ml/min]	Heating-process	Hot tool temp. T_H [^{0}C]	Matching-pressure p_A [N/mm^2]	Heating-time t_H [s]	Joining pressure p_j [N/mm^2]	Joining, cooling time [s]
PE-HD	0.1–20	Contact (con-ven-tional)	160–280*	0.2–0.5	10–60	0.05–0.25	10–20
PE-LD	20–60		160–280*	0.2–0.5	15–60	0.05–0.15	10–20
(MVR 190/5)	0.1–60	Non-contact	400–550	–	5–15	0.05–0.25	10–20
PP	0.3–3	Contact (con-ven-tional)	200–270	0.2–0.5	10–60	0.05–0.25	10–20
PP/ EPDM (MVR	3–40		180–220	0.2–0.5	15–60	0.05–0.15	10–20
230/2.16)	0.3–40	High-temp*.	300–400	0.1–0.5	3–10	0.05–0.25	10–20
	0.3–40	Non-contact	400–550	–	5–15	0.05–0.25	10–20

*) By heated tool temperature above 270 °C; exclusively heated tool without PTFE-coating

3.2.2.1   High Temperature Welding

During hot plate welding of some thermoplastics, sticking to the hot plate becomes a problem. When using higher hot plate temperatures with these thermoplastics, the PTFE non-stick coating cannot be used. This is especially a problem with low surface energy thermoplastics, whose "melt-strength" and viscosity are low. Because of their low surface energy, these thermoplastics easily stick to the hot plate. When retracting the weld part from the hot plate, the melt easily tears resulting in sticking and stringing of the melt. The melt that remains stuck to the hot plate can then degrade and transfer to subsequent welds resulting in poor and aesthetically unappealing welds.

To avoid this, high temperature welding uses hot plates without anti-adhesive PTFE coating. Depending on the type of plastic, the hot plate surface temperatures range between 300 and 400 °C. The process sequence is identical to conventional hot plate welding. Times for matching and heating phase are extremely short – only between 2 and 5 s. Viscosity of the melt decreases strongly due to the high temperatures and the short heating times. The lowest viscosity is at the contact faces between welding part and hot plate. This is the point where the melt will peel off the hot plate when removing the welding parts. The residual material that remains on the hot plate surface then evaporates or oxidizes prior the next heating operation resulting in a clean hot plate for the next cycle. Due to these extreme hot plate temperatures, thermal degradation of the melt can be expected – in spite of short matching and heating times. Usually, the degraded material will be largely forced out into the bead during joining, and the weld quality will only be slightly affected. Principally, however, reduced weld strengths have to be expected with high temperature welding, and it cannot be applied to all thermoplastics. With some types of thermoplastics the residual materials do not completely evaporate or oxidize. Instead, they accumulate over time and prevent proper welding. In current industry practice, good results can be expected with high temperature hot plate welding with the following materials:

- PP and PP copolymers (e.g., for welding automotive batteries)
- ABS and PMMA (e.g., for welding automotive rear lights)

In the high temperature welding of reinforced, filled, or other special types of plastic, residues on the hot plate do not evaporate or oxidize completely. In those cases, cleaning devices should be installed that automatically remove or clean the hot plate between welds. In addition, fume extraction devices should be installed above the machine to remove the vaporized material.

3.2.2.2   Non-Contact Hot Plate Welding

Another way to avoid sticking of polymers to the hot plate is using non-contact heating. In this case, the joining surfaces are melted without direct physical contact with the hot plate but rather through radiant or convection heating. The hot plates' surface temperature is raised to high levels of 400 to 550 °C and the parts are placed at a distance of 0.5 to 1 mm from the hot plate. In order to ensure uniform heating of the joining surfaces, the distances between joining surfaces and hot plate have to be as uniform as possible. Deviation from plane parallelism between the parts and the hot plate should not exceed a value of 0.2 mm. Hot plate temperatures and the heating time should be selected in such a way that the joining surfaces are sufficiently plasticized, but do not suffer any thermal degradation. Following the heating phase, the other phases of the non-contact hot plate welding process are identical to conventional hot plate welding.

Maintaining a constant gap or the desired parallelism between the parts and the hot plate is possible in practice only for small parts. For large parts, molding warpage and cavity-to-cavity variations are difficult to control to such a high precision. Therefore, non-contact hot plate welding remains restricted to moldings that are produced true to design dimensions. Usual component sizes do not exceed 100 x 100 mm. It should be noted that non-uniform heating could also result from convection heating of the air between the faying surface and the hot plate. This is often called the "chimney effect." Using a heated tool that is placed horizontally instead of vertically can reduce this undesired effect.

3.2.3   Welding Parameters

Experimental and theoretical research has shown that the following parameters influence the quality of a hot plate weld:

- Hot plate temperature
- Matching time
- Matching pressure/displacement
- Heating time
- Heating pressure
- Change-over time
- Joining displacement
- Joining pressure
- Cooling time

Below is a short description of these parameters:

Hot Plate Temperature

The hot plate temperature is the temperature on the working surface of the hot plate. It must be adjusted in conformance with the selected heating method and the heating conditions of the material to be welded such as melting temperature, melt viscosity, and degradation time-temperature limits. Conventional hot plate welding usually works with hot plate temperatures in the range of 30 to 100 °C above the melt point of the thermoplastic. The upper temperature is determined by the degradation temperature (and heating time) and the self-ignition temperature of the plastic. High temperature welding operates with hot plate temperatures, which range above the decomposition temperature of the thermoplastic material, preferably 100 to 200 °C above the melt point. In non-contact heating, the temperatures lie 300 to 400 °C above the melt point of the plastic. The radiant energy given off by the hot plate depends not only on the hot plate temperature, but also on the surface color of the hot plate (or more precisely the hot plate material's emissivity). The darker the surface, the higher the energy radiated.

Matching Time/Displacement Time

Matching time refers to the time that the parts are heated under high pressure to melt and squeeze out warpage or dimensional unevennesses, so that the joining surfaces are in full contact with (conform to) the hot plate.

Matching Pressure/Displacement

Matching pressure is the pressure used to press the weld parts against the hot plate during the matching phase. While a relatively high pressure is used during the matching phase, it should not be so high as to cause excessive deformation of the parts. In some cases, instead of applying pressure for a predetermined time, a mechanical stop is used to impose a matching displacement.

Heating Time

Heating time is the time during which the joining part is in contact with the hot plate with very low or zero pressure. For contact and high temperature welding, heat conducts from the hot plate to the parts. For non-contact heating, heat is transferred to the parts by radiation and convection. The correct heating time can be found either theoretically or experimentally.

Heating Pressure

The heating pressure is the pressure that is applied to ensure good contact between the parts and the hot plate during the entire heating time. In case of excessive heating pressure, the melt will be pressed into the flash, and there will be insufficient amount of melted material. Often mechanical stops are employed to prevent squeezing the entire melt out of the bond-line, which could result in a "cold weld". In case of too little pressure, full contact and thus good heat conduction between hot plate and joining part might not be ensured.

Change-Over Time

Change-over time is the time required to remove the parts from the hot plate, retract the hot plate, and bring the parts together for joining. Change-over time should be as short as possible, in order to avoid surface cooling of the melted material. For properly designed welding machines it is no problem to keep change-over times as low as 2 to 3 s, even for moldings with a size of about 400 x 400 mm.

Joining Pressure

The joining pressure is used to press the parts together during the joining or forging phase. The joining pressure must be set to ensure that the correct amount of melt is squeezed into the flash. If the joining pressure is too low, the squeeze-out will not be sufficient to remove entrapped air and facilitate intimate contact at the interface. If the joining pressure is too high, the entire amount of melt will be squeezed out of the joining area. Usually, mechanical stops are employed to prevent squeezing the entire melt out of the bond-line, which could result in a "cold weld". In this case, the material layers are below the melt point when they meet resulting in a weak tack weld. The joining pressure is maintained during the entire cooling time. However, if mechanical stops are employed, the joining pressure at the bond-line may vary as the force is transferred to the stops.

Joining Distance

The distance the joining parts travel during the joining operation.

Cooling Time

Cooling time is the amount of time between joining and solidification of the melt, i.e., the moment when the temperature of semi-crystalline thermoplastics falls below the crystalline melting point or when the temperature of amorphous thermoplastics falls below their glass transition temperature. The welding parts can now be removed from the machine. The welded moldings should not be subjected to any stress during removal and subsequent storage, until they have cooled all the way down to room temperature.

Each of the above described welding parameters is of essential importance for the weld quality. However, they should not be considered individually, rather their mutual influence is to be taken into account.

3.2.4   Determination of Melt Layer Thickness

The quality of the weld depends on the melt layer thickness that is generated during heating and the fraction of the melt layer that has been squeezed out during joining. Figure 3.3 shows an example for determination of the melt layer thickness and of the remaining melt layer thickness for HDPE components with a wall thickness of d = 2 to 4 mm. It was found that for HDPE, heating must be sufficiently long to form a melt layer thickness (L_o) that is 34% of the wall thickness. During joining approximately 75% of the

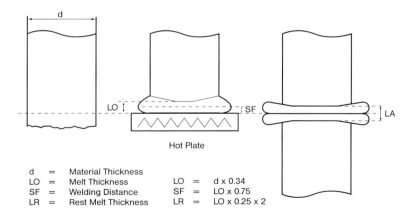

Figure 3.3 Desired melt layer thickness and squeeze-out for HDPE parts with wall thicknesses between 2 and 4 mm

melt layer needs to be squeezed out to achieve optimum weld quality. Therefore, at the end of joining the remaining melt layer thickness (L_R) is 25% of L_o. For larger wall thickness, the melt layer thickness during heating is a slightly smaller fraction of the wall thickness. These melt layer thickness values should be used as guidelines. Specific melt layer thickness values should be developed from experiments that correlate melt layer thickness during heating and remaining melt layer thickness after joining to weld strength.

The melt layer thickness that develops during heating can be measured in a variety of ways. For some semi-crystalline polymers, like HDPE, the molten material is transparent and it can easily be visualized for measurement purposes. Alternatively, microscopic examination of microtome slices can usually be used to identify melted regions. Therefore, to measure the melt layer thickness at the end of heating, the parts should be removed from the welder prior to change-over and be allowed to cool for microtome slicing. Measurement of the squeeze-out during joining can be done by measuring the length of the parts prior to joining and then after joining.

3.3   Quality Control

Customers ordering plastic components may require evidence of weld quality – especially in cases when parts are highly stressed or they are important safety components. There is no non-destructive test method known which would offer perfect quality assurance. Therefore, it is recommended to include quality control systems as part of the operating sequence. In hot plate welding, quality influencing parameters (e.g., temperature, pressure, and time) can be precisely controlled, which provides a good basis for the use of various control methods

3.3.1   Control of Processing Parameters

This is the easiest and most frequently used method of quality control. The parameters ensuring good welds are permitted to vary within an admissible tolerance range and they are recorded during each weld cycle. As soon as one of the parameters is outside the tolerance range, the machine stops and alerts the operator. At that time corrective action can be taken and welding can continue.

3.3.2   Statistical Process Control (SPC)

Statistical process control constitutes a further development of parameter control. SPC is used to monitor the process parameters and if needed to control them within a preset range. However, application of SPC to manufacturing processes is not always possible because of the interaction between process parameters, which classical SPC can not readily account for.

3.3.3   Continuous Process Control

In contrast to the spot-checking of process parameters, as is the case with classic SPC, the process parameters can be monitored and controlled continuously. For this purpose, a statistical process model that may be based on a suitable database of experimental tests is prepared. The process parameters are then compared to the database with regard to one or more quality features through a step-by-step regression process.

3.4   Equipment Description

Hot plate welding equipment is divided into standard machines and custom machines. *Standard machines* can be used to weld a variety of parts of different geometries by interchanging fixtures and hot plates. *Custom machines* are distinguished by the fact that they are predominantly developed and used for a specific part. In addition to welding, custom machines frequently include machining, feeding, and component removal devices. Both standard and custom machines include timing control of the operating sequence to ensure reproducible working conditions and constant weld quality. With semi-automatic machines, the joining components are loaded and unloaded manually by an operator, while with fully automated systems part loading and unloading are also automated.

Both types of welding machines are usually operated pneumatically. Bigger machines with heavy hot plates, working carriages, and toolings are operated hydraulically. Newer machines are equipped with electromechanical motion control. The electromechanical designs allows for quick change-over times through fast motions but they also allow the final stage of the motion to be slowed down so that the welding components will not hit the hot plate, which could result in accelerated wear of the anti-adhesive coating. Also, the melted surfaces of both joining components must not be brought together at high speed, or melted material would be forced out of the joining zone, which again would result in poor weld quality.

3.4.1   Sequence of Operation of Semi-Automatic Welding Machines

Semi-automatic machines are available on the market for welding parts, which may range in size from 20 x 300 mm (joined in multiple-cavity tools) to 1000 x 1200 mm (such as transportation pallets). A distinction is made between horizontally and vertically operating machines, indicating how the moldings move towards each other. Normally, longer moldings are welded horizontally (e.g., pipes, profiles), while moldings with internal fittings (e.g., starter batteries) are welded vertically. Figure 3.4 shows a semi-automatic welding machine with vertical part motion and Fig. 3.5 shows a semi-automatic welding machine with horizontal motion.

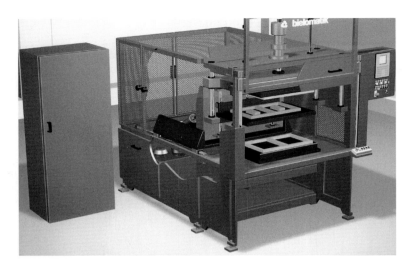

Figure 3.4 Semi-automatic vertical hot plate welding machine

Figure 3.5 Semi-automatic horizontal hot plate welding machine

Semi-automatic hot plate machines operate in the following sequence:

• All working units carrying holding fixtures and hot plate are in the initial position. The guard door at the machine's operating side is open. The operator places the welding components in the holding fixtures and checks if they are in the correct position. In vertically operating machines, the bottom components are usually placed in the bottom holding fixtures and the top components are placed on top of them. Quite often, positioning aids are necessary.

- The operator closes the guard door, and initiates the automatic cycle of the machine.
- The welding components are clamped in their fixtures. For vertical machines, the upper fixture moves down, picks up and clamps the welding component and then returns to the upper position.
- The hot plate moves to the working position and both working units carrying the welding components move against the hot plate. At this time the matching phase begins as the parts are pressed against the hot plate. If mechanical stops are installed to limit the squeeze-out, then the carriages will come in contact with them at the end of the matching phase. If limits are set by pressure-control, at the end of a preset time the pressure is released and the welding components remain in contact with the hot plate without pressure.
- After heating time has elapsed – set on respective timers – the carriages return to their initial position and then the hot plate is retracted to its initial position.
- Joining now takes place as the carriages move towards each other as quickly as possible – with reduced speed shortly before the parts come in contact. Pressure is maintained while the weld interface cools. Mechanical stops may be used to limit the squeeze-out of the melt during the joining phase.
- After cooling is completed – set again on respective timers – one side of the welding component is released from its holding fixture. In vertical machines, this will be the upper tool. The carriages return to initial position, the second holding tool is released and the guard door opens. The operator can remove the welded part(s) from the machine.

3.4.2   Sequence of Operations of Fully Automatic Machines

The operations of fully automatic welding machines include, in addition to hot plate welding, loading, machining, and possibly welding of some components using other processes (e.g. spin, vibration, and ultrasonic welding). The sequence for hot plate welding is identical to the one for semi-automated machines except that loading and unloading is done automatically using robots or other means and the welding cycle is initiated automatically. Figure 3.6 shows a fully automated machine designed to machine and to weld automotive fuel tanks.

- The production begins with robots loading the blow-molded tanks onto special conveyor pallets, where they are clamped. These pallets run automatically through the stations of the production equipment.
- The tank openings are cut out in the first working stations. The cut-out pieces are automatically removed, ejected, and counted.
- Functional components are welded to the tank in the subsequent stations. Due to the large amount of components to be welded – up to 34 parts – several welding stations are necessary. Hot plate welding has wide operating windows making it the preferred

process for welding automotive fuel tanks and components because of safety requirements. The components to be welded to the tank are automatically fed in from maga-
zines or vibrator pots. They are seized by the welding stations and joined. Each
component is joined in a separate and individually operating welding station. The control system for each welding station must be designed to allow for different matching
times, to correct for part warpage, or uneven weld surfaces.

- Welding of additional components to the fuel tank system could also be done using
  other processes such as spin or ultrasonic welding.

- At the end of the production line, the pallets holding the tank are lowered below the
  main machining and welding area and they are transported to the initial position, where
  they rise to the top level again to enable release of the completed fuel tank and loading
  of another pre-assembled fuel tank on the pallet.

Figure 3.6 *Fully automated machine for production of automotive fuel tanks*

The following key components can be found in each hot plate welding machine:

Hot Plates

Hot plates are the heart of the welding unit. Correct dimensioning and heat capacity as well
as uniform temperature distribution on their surface are of vital importance for the quality
of the weld. Hot plates are made from different materials – depending on the welding
variant used – see below for detailed information. Hot plates are heated exclusively by
electric heating elements, which must be placed appropriately to ensure uniform temperature distribution. As the hot plate size increases, so will the non-uniformities in temperature. Below are general guidelines showing acceptable temperature variation in hot plates
of different sizes:

- Diagonal dimension up to 350 mm: 10 °C
- Diagonal dimension up to 700 mm: 14 °C
- Greater than 700 mm: 18 °C

Depending on the thermal mass of the parts, when they contact the hot plate they will lower its temperature as heat is conducted away from the hot plate and into the parts. The hot plate should have sufficient thermal mass so that the temperature drop on the hot plate surface during the matching and heating phases does not exceed 10 °C. The temperature is maintained constant by means of electronic temperature controllers, which usually provide a sufficiently constant temperature.

Hot Plates for Conventional Hot Plate Welding

Hot plates must be designed for a working temperature of at least 270 °C. They are predominantly made from aluminum alloys, i.e., from a material that conducts heat very easily and is resistant to corrosion. To avoid sticking of polymer melt to the hot plate it is usually coated with a PTFE layer that is bonded to the metal plate. For flat hot plates, it is common to cover them with PTFE-coated glass fiber fabrics, as they can be easily changed. Good results were achieved with glass fiber reinforced PTFE films with a thickness of 250 μm. When setting the hot plate temperature, it is important to note that the PTFE films act as thermal insulators and they decrease the surface temperature of the hot plate by about 20 °C.

For parts with complex and three dimensional weld surfaces, flat hot plates with inter-changeable fixtures are used (see Fig 3.7). The complex fixtures are PTFE-coated with a layer that is 30 to 50 μm thick. Here too, the temperature insulating characteristics of the coatings have to be considered – usually 5 to 10 °C.

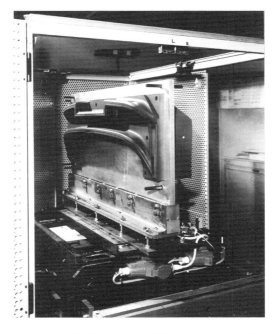

Figure 3.7 *Flat hot plate with complex interchangeable fixtures mounted on it*

PTFE-coatings need to be renewed from time to time. Their lifetime depends on the hot plate temperature and on the evenness of the components' contact surface. Burrs and elevated spots on the contact surface of the welding components can destroy the PTFE coating quickly. To facilitate smooth, uninterrupted production it is recommend to have 3 sets of interchangeable fixtures available to allow quick exchange when re-coating is required. PTFE can sustain maximum operating temperatures of 270 °C, which limits the useful operating temperature range of the hot plate to 270 °C. In spite of intensive research, no other non-stick coatings are available. The use of mold release is not recommended because it may adversely affect the weld strength.

Hot Plates for High-Temperature Welding

These can be used up to a maximum working temperature of 430 °C. Since aluminum anneals at these high temperatures, aluminum bronze alloys are used. Otherwise, the hot plates and interchangeable fixtures for complex parts have a similar design as those used for contact heating. Aluminum bronze alloys have a lower thermal conductivity than aluminum; therefore special care must be taken in designing hot plates for high temperature welding to ensure that the temperature is uniform along the surface of the hot plate. Since non-stick coatings cannot be used due to the high operating temperature, it is important to have smooth surfaces in contact with the weld components. When welding PP, chemical reactions may occur with copper in the hot plates, leading to pitting in the interchangeable fixtures or the hot plates. Therefore, these surfaces should be re-finished from time to time.

Hot Plates for Non-Contact Welding

These hot plates are used for working temperatures up to 550 °C and they are made from aluminum bronze or stainless steel. In current industry practice, only hot plates with flat surfaces are used.

Clamping Fixtures

The clamping fixtures, which are attached to the carriage on the welding machine, are used to hold the parts during the welding process. Furthermore, they must provide uniform support to the parts to avoid deformation under welding pressures. Depending on the shape of the components, different types of fixtures are possible. Preferably, mechanical clamping fixtures should be used which are activated either pneumatically or hydraulically. Vacuum suction equipment, which provides uniform support to the parts, is also suitable. If components with delicate surfaces have to be welded, the contact surface of the fixtures can be plastic-coated to avoid scratching. If a large number of small parts have to be welded, multiple-cavity tools for simultaneous welding of two or more components can be used to increase efficiency (see Fig. 3.8).

If the clamping fixtures have to be changed frequently, it is recommended to integrate the mechanical stops for the matching and joining phases into the fixtures and hot plates. Clamping fixtures that are used with high-temperature or non-contact hot plate welding may require air or water cooling channels to avoid excessive heating.

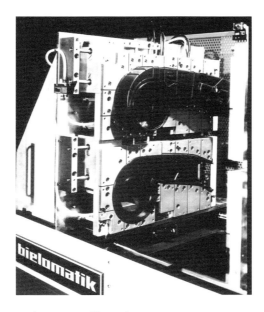

Figure 3.8 *Fixtures for simultaneous welding of two components*

3.5   **Joint and Part Design**

The joining components as well as the joint itself must be designed to be suitable for welding and they must also be made from suitable materials. The design requirements for the joining components include:

- Design of the joining components must allow accurate fixturing in the welding machine
- Joining components must have identical sized and shaped weld areas
- Joining components must be sufficiently rigid to withstand the pressures applied during the matching and joining phases
- Clamping fixtures should be able to contact and support the parts as close to the joining surfaces as possible to avoid misalignments
- Sufficiently large radii at corners and transitions to reduce stress concentration
- Joining surfaces should be as planar and flat as possible
- Avoid injection molding weld-lines (knit lines) in the joining area.
- For non-contact heating, use raised joining surfaces with greatest possible distance away from the rest of the part to avoid heating regions away from the joint area.

When considering joint design it is important to remember that a weld bead (or flash) is necessary to achieve good joint quality. The bead occurs as a result of the squeeze flow, which is necessary to remove entrapped gases from the weld interface and to achieve intimate contact for intermolecular diffusion. Usually the weld bead is uniform and homogenous – except for glass or carbon fiber filled materials – and it does not affect the performance of the welded parts. There may be applications where the weld bead can adversely affect the performance of the part. For example, in medical applications, the weld bead (flash) can affect the flow of a fluid such as blood resulting in damage or separation. In other instances, the weld bead is not desirable because of aesthetic reasons. Therefore, a variety of joint designs are available including ones that have flash traps to avoid adverse affects of the weld bead.

Simple Butt Joint

The simple butt joint is the most common joint design for moldings (see Fig. 3.9).

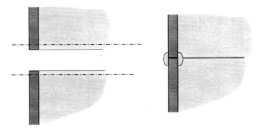

Figure 3.9 *Simple butt joint design*

Modified Butt Joint

This joint design covers the bead on both sides making it less visible (see Fig. 3.10). The back of the rib can be used to support the joining parts in the clamping tool.

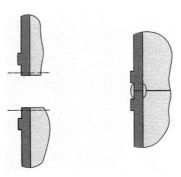

Figure 3.10 *Modified butt joint design*

Enlarged Joining Surface

When welding thermoplastics with large amounts of fillers, the weld strength is lower than the bulk material strength. Load transfer through the joint can be increased by enlarging the joint surfaces (see Fig. 3.11). Here again, the back of the surrounding rib can be used to fix the joining parts in the clamping tool.

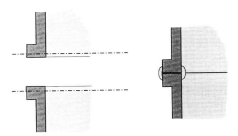

Figure 3.11 *Enlarged joint surface design*

Butt Joint with Internal Flash Trap

The weld bead is hidden on one side by a cover rib or flash trap for aesthetic reasons (see Fig. 3.12). The surrounding groove which will take up the bead should be at least 1 mm wide (depending on the bead volume).

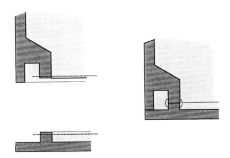

Figure 3.12 *Butt joint with internal flash trap*

Butt Joint with External Flash Trap

Joints with a single flash trap can also be designed to hide the flash on the outside (see Fig. 3.13). This joint design is good for joining vent housings in order to prevent noises from the incoming air.

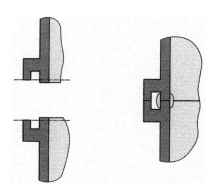

Figure 3.13 *Butt joint with external flash trap*

Butt Joint with Double Flash Trap

The flash can be hidden on both sides of the joint for aesthetic requirements (see Fig. 3.14). This joint design is good for joining household appliances.

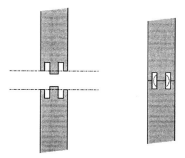

Figure 3.14 *Butt joint with double flash trap*

Butt Joint with Flash Cover

The weld bead can also be covered by a thin rib (see Fig. 3.15). The distance between joint surface and the inside of the flash cover should be at least 1 mm to allow room for the flash. This joint design is good for sealing automotive batteries.

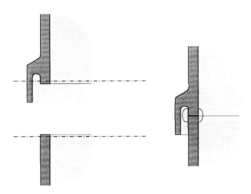

Figure 3.15 *Butt joint with flash cover*

Groove Butt Joint

This weld design is used for welding injection-molded components to extruded containers (e.g., automotive fuel tanks), see Fig. 3.16. The joint design is not perfect, neither with regard to heating (quite often, there is more time needed to heat and melt the molded part than for the extruded part) nor with regard to the melt flow during joining. However, other solutions are not practical.

Figure 3.16 *Groove butt joint design*

Usually, removal of the weld bead is not necessary. However, in special cases, the bead may need to be re-formed or removed for either functional or aesthetic reasons. Figure 3.17 shows an approach for re-forming the weld bead while it is warm to make it flush with the surface. For small and rigid injection molded parts, the welder can be equipped with mandrels or rings that force the weld bead into a groove while it is hot. The weld bead cools off under the forming tool and if bead and groove are correctly sized, it results in a flush almost bead-free surface. An alternative approach is to cut or pinch off the bead using a knife after the parts have completely cooled (see Fig. 3.18). With appropriate tools this approach can be used only on small rigid parts.

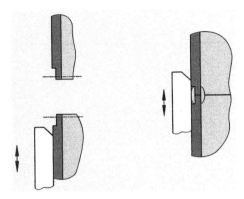

Figure 3.17 Re-forming of weld bead to make it flush with surface

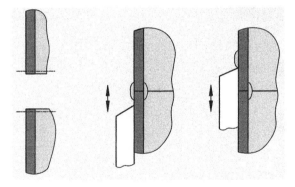

Figure 3.18 Cutting of weld bead with a knife

3.6   Applications

Applications for hot plate welding were divided into two major groups: production applications and pipe welding. Both of these areas are of major importance, yet they differ in terms of equipment and part and joint configurations.

3.6.1   Production Applications

Welding of Barrels Made from HDPE

Extruded pipes of approx. 500 mm diameter are cut to barrel height. Bottom and lid are welded to the barrel in one joining operation on a special-design hot plate welding machine. Since the lids are injection molded, the welding process must be robust enough to accommodate the difference in melt viscosity between the extruded pipe and the injection molded lids. In this case, dual hot plates were used to heat the lids and pipe to different temperatures to achieve similar viscosity of the melt. Since these barrels are used to transport chemicals, high long-term weld strength was required. Conventional contact hot plate welding with simple butt welds was used. Due to a wall thickness of 6 mm, the weld area was large, requiring high forces; therefore hydraulically operated machines were used. The cycle time including loading and unloading of the parts was 70 s. Figure 3.19 shows the assembled barrel.

Figure 3.19 *Hot plate welding of lids to pipe to fabricate HDPE barrel*

Welding of Transport Pallets Made from PP

Plastic transport pallets have a standard size of up to 1000 x 1200 mm and are made of two injection-molded halves which are joined by hot plate welding (see Fig. 3.20). They are welded lengthwise, by means of 9 raised ribs. Only conventional contact hot plate welding is able to meet the strength requirements. The pallets were welded on a vertical hot plate welder using a simple butt weld. Because of the large area, high forces were needed to apply the necessary pressure. Therefore, a hydraulically operated machine was used. Because of the size of the parts, warpage was a problem and high matching times were required. Therefore, cycles times were about 90 s.

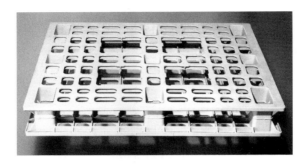

Figure 3.20 *Hot plate welded PP transport pallets*

Air Intake Pipe Made from Talcum-Filled PP

Automotive air intake pipes are made from two injection molded halves that are joined together (see Fig. 3.21). To save weight and material costs, the walls are very thin, which results in warpage. Therefore, the design of the clamping fixtures was important, so that they would hold the welding parts in correct welding position. To avoid noise by incoming air, a butt weld with interior flash trap and enlarged joint area was used. Due to the talcum reinforcements, it was necessary to enlarge the joint area and increase the load-carrying ability of the joint to be similar to that of the wall. Conventional contact hot plate welding was used with a cycle time of approximately 40 s.

Figure 3.21 *Hot plate welded talcum-filled PP air intake pipe*

Fuel Tanks

Fuel tanks are made of HDPE and they are generally blow-molded. Then holes are machined into the fuel tank and hot plate welding is used to attach important functional parts to the tank. Complicated fuel tanks may have up to 34 components attached. The matching time may be different for each component to overcome warpage in the tank or the part to have even surfaces for welding. Each welding component has an individual welding station, which is operated independently. For joining, groove welds were chosen, as raised ribs cannot be provided at the tank top. The cycle times for welding each part on fully automatic welding machines were less than one minute. Figure 3.22 shows the blow molded fuel tank and Fig. 3.23 shows all the parts that are hot plate welded to the tank.

Figure 3.22 Blow molded HDPE fuel tank

Figure 3.23 Functional components that are hot plate welded to the blow molded fuel tank in Fig. 3.22

Automotive Batteries

The cases and lids of automotive batteries are usually made of PP-copolymers. After the lead cells are inserted, connected and tested, the battery lid is welded to the battery case. High temperature welding was used because the walls are thin and the polymer has a low melt viscosity allowing for ample squeeze flow. Smoke extraction devices had to be installed above the machine. The joint design was butt joint with a flash cover. The welding machines are able to seal two batteries at a time with a cycle time of less than 30 s. Figure 3.24 shows the assembled battery.

Figure 3.24 *Hot plate welded PP-copolymer automotive battery*

Tail Lights

The housing of automotive tail lights is usually made from acrylonitrile-butadiene-styrene (ABS), while the lenses are made from polymethyl methacrylate (PMMA) (see Fig. 3.25). In special cases, the lenses may be made from polycarbonate (PC). In this case, either conventional or high temperature welding can be used. When using a material combination of ABS and PMMA, the welding temperatures are similar and a single hot plate can be used. When welding ABS and PC, different welding temperatures are required and the use of dual hot plates is recommended. The vertical welding machines normally are equipped with double clamping fixtures enabling welding a right-side and left-side lights simultaneously. A modified butt joint was used. With a hot plate temperature of 370 °C the cycle time was about 60 s for one pair of tail lights.

Figure 3.25 Hot plate welding of automotive tail light

Automotive Center Console Made from ABS

The center console is made of ABS. It requires precise alignment of the parts relative to each other for aesthetic reasons. Since ABS can absorb moisture, welding has to be done immediately after injection molding. Depending on the type of polymer used, conventional hot plate welding as well as high-temperature welding is applicable. A standard hot plate welding machine was modified to be able to weld all three parts in one operation. Figure 3.26 shows the parts of the center consol prior to welding.

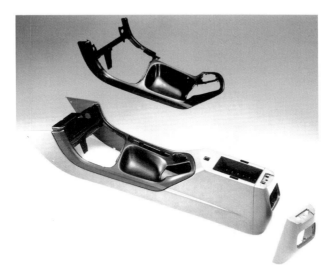

Figure 3.26 Center consol components

Carburetor Floats

Carburetor floats are made from polyoxymethylene (POM) because of its good resistance to gasoline. POM has a very low melt viscosity and it is not suited for conventional hot plate welding. Since these floats are very small and true to size and shape, non-contact hot plate welding was used. The hot plate was heated to a temperature of approx. 500 °C with a gap of about 1 mm between the parts and the hot plate. To increase production speed, the welding machine was equipped with 4-cavity tooling. When the floats were sealed, a metal part was heat-staked to the float at the same time. The clamping fixtures were cooled to avoid excessive heating of the parts away from the joint area. The components were placed in the clamping fixtures by means of a loading tool, which was loaded by the operator during the welding. After welding, the welded parts were ejected and dropped out of the machine, allowing the loading tool to come in and load the next batch. Cycle time for 4 parts was approx. 40 s. Figure 3.27 shows the welding parts and Fig. 3.28 shows the multiple-cavity tooling.

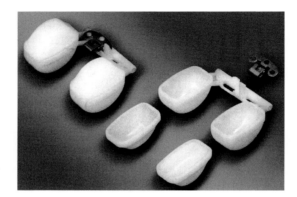

Figure 3.27 *Hot plate welded POM carburetor floats*

Figure 3.28 *Clamping fixtures for POM carburetor floats from Fig. 3.27*

3.6.2   Hot Plate Welding of Plastic Pipes

One of the most common applications of the heated tool (hot plate) welding method is joining of plastic pipes. In many industries it is commonly called fusion welding, and depending on the joint configuration and the specific tool used, it can be referred to as butt fusion, saddle fusion, or socket fusion welding.

One of the main reasons this application of hot plate welding is becoming more popular in the US and has been for many years in other countries, are safety issues related to earthquakes. Compared to steel pipes, plastic pipes are able to deform during an earthquake rather than rupture.

While the basic welding process is essentially the same as described in joining of molded parts in mass production (see Section 3.2), the pipe welding process has its own important specifics, related to special joint configurations: butt, saddle, and socket joints, and their associated welding procedures.

While there is a wide assortment of plastics that are hot plate welded in industrial applications, most of the pipes joined by welding are made from a limited number of thermoplastic materials: polyethylene, polypropylene, polybutylene, and polyvinylidene fluoride. The differences in physical properties of each of these materials may be significant enough to affect the joining procedures.

The equipment commonly used for pipe welding is also highly specialized and differs conceptually from the equipment used for volume production. Portability and ability to adopt a range of pipe diameters are necessary features of field welding equipment that varies widely in degree of automation (from manually operated to fully automated machines).

Environmental conditions that may vary widely also add special requirements for the joining procedure and equipment design and have to be taken into account. The presence of dust and other contaminants during welding in the field requires special surface preparation. Factors like ambient temperature and wind conditions affect process parameter selections.

Fusion welding of pipes includes three different processes:

- Butt fusion
- Socket fusion
- Saddle/Sidewall fusion

3.6.2.1   Butt Fusion

Butt fusion is one of the most widely used pipe-joining techniques. Its applications range from large diameter polyethylene pipes in a variety of distribution and drainage systems to pipelines in industrial and municipal buildings, including systems that require welding in a clean-room environment. The technique precludes the need for specially modified pipe ends or couplings and produces strong, economical and flow-efficient connections. It can be used to weld pipes directly one to another or to appropriate fittings (e.g., bends, tees, etc).

In-field and in-shop butt fusion welds may be made readily by trained operators using specially designed butt fusion machines that clamp and precisely align the pipe ends for the fusion process. The machines are equipped with a flat hot plate and a planer unit, which is used for mechanically facing the mating surfaces of the pipes. Butt fusion welding machines clamp the pipes in carriages that provide longitudinal movements and force on the pipe ends (see Fig. 3.29). The wide variety of butt fusion equipment reflects the breadth and diversity of requirements from different industries that use this joining method in construction of specialized pipe systems. When smaller size pipes are being joined, inexpensive butt fusion machines with manually moveable clamps are still commonly used in many cases. For high-end applications, programmable machines with electromechanical drives and precise force or displacement control are also available.

Figure 3.29 *Manually operated bench top machine for fabricating segmented bends*

While most equipment vendors still prefer to design equipment that operates in a force control mode, some equipment makers offer butt fusion machines with a force-controlled, displacement-monitored capabilities, or with a displacement limit control. Many of these advanced machines are used for joining pipes in a clean-room environment – one of the most challenging and fast-growing applications of butt fusion technology. Another trend brought about by the industries using butt fusion in a clean-room environment are welds with reduced bead made by means of non-contact hot plate heating with welding distance control and convection cooling of the joint, that results in reduction of the cooling time while ensuring consistent weld quality.

Butt fusion machines for larger diameter pipes usually utilize hydraulic power drives. These machines are robust, easy to operate and are designed to accommodate a specific range of pipe diameters up to 1200 mm either in a workshop or in a building-site (in-field) condition. Figures 3.30–3.32 show some examples of stationary and building-site machines.

Portable machines are commonly available on a frame or on a cart configuration and have pipe drag compensation features that allow automatic control of the pressure necessary for facing, heating, and fusion cycles regardless of amount of pipe drag. Most of the modern machines are partially or fully automated; some of them are equipped with CNC control that allows continuous control of welding distance, force, and temperature and they provide full logs of all parameters for every completed joint.

Figure 3.30 CNC-controlled 550 mm field machine with hydraulic power drive

Figure 3.31 Semi-stationary 225 mm machine

Figure 3.32 CNC controlled stationary machine for prefabrication of elbows, T-pieces, and Y-pieces

Regardless of the equipment and conditions in which the joint is made, pipe butt fusion procedure can be arbitrarily divided into six main steps:

Loading of parts

Components have to be aligned and secured in position so they will not move unless they are moved by the clamping device. While the following operations can be mechanized or fully automated, this initial step is usually performed manually, regardless of the degree of automation of the welding machine.

Joint Preparation

The pipe ends are faced to establish clean, parallel mating surfaces. Most equipment manufacturers have incorporated the rotating planer block design in their facers to accomplish this goal. The operation provides for a perfectly square face, perpendicular to the centerline on each pipe end and with no detectable gap between mating surfaces when they are brought together to verify the quality of surface preparation. Figure 3.33 shows a typical setup for this phase.

Planer

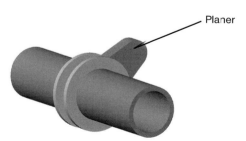

Figure 3.33 Planer used to square the pipe ends

The pipes profiles are rounded and aligned with each other to minimize mismatch (high-low) of the pipes walls. This is accomplished by adjusting the clamping jaws until the outside diameters of the pipe ends match.

Melt Down Phase

The pipe faces are heated and melted on a heated plate under recommended parameters for temperature, time, and interfacial pressure. During the heating phase, the heat will penetrate into the pipe ends and create a layer of molten material on the mating surfaces so that an initial bead will be formed around the pipe edges. Some manufacturers recommend to carry out the heating phase in two stages, matching (with adjustment pressure) and heating (without, or with minimum pressure), but many others use a one-phase heating cycle. A single hot plate with anti-adhesive coating is used to simultaneously heat both joining surfaces. They are normally equipped with suspension and alignment guides that center them on the pipe ends.

The development of uniform melt layers on mating surfaces of pipes made from similar materials may be readily accomplished with existing equipment. The same task is more difficult when fusing materials with different melting characteristics, because commonly available pipe welding machines are not equipped with dual hot plates. As described in Section 3.7.6, uniform melt layers on both pipes can be achieved by heating of the lower melt index material earlier than the higher melt index material so that both materials emerge from the hot plate with relatively uniform melt viscosities. This technique, used by some installers of PE piping, was subjected to an evaluation by the Plastic Pipe Institute, who found that this approach yields satisfactory joints. Figure 3.34 shows a typical setup for this phase.

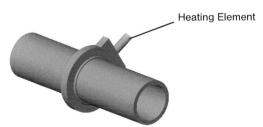

Heating Element

Figure 3.34 *Matching and heating phase*

Change-Over and Fusion Phase

After the pipe ends have been heated for an appropriate time, the hot plate is removed and the molten pipe ends are brought together with the recommended pressure. During this phase, the melt is squeezed out from the melt zone to form a bead inside and outside of the pipe surface. This way, any possible contaminations will be displaced from the joint. Most of the guidelines specify the interfacial pressure dependent on the material melt flow index. Fully automated machines can also apply pressure until a predetermined displacement is reached. In some industries, the bead size is used as a guide for applying interfacial pressure in manually operated equipment. Figure 3.35 shows a typical setup during this phase.

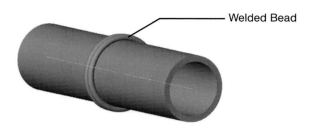

Figure 3.35 *Controlling joint pressure based on welded bead size*

Hold Phase

The joint is cooled under pressure until the molten material is fully solidified. The proper cooling time for the joint is dependent on the material, the pipe diameter, and the wall thickness. In most of automated equipment the cooling time is preprogrammed for the specific type of pipe, and some machines are equipped with automatic locking devices that assist the operators to accomplish this step.

In some pipe system application, the bead from the butt fusion process may be undesirable and should be removed to provide smooth internal and external pipe surfaces. The inner surface of the removed bead may also be inspected to get an indication of weld quality.

3.6.2.2   Socket Fusion

This specialized technique consists of simultaneously heating both, the external surface of the pipe and the internal surface of the socket fitting by a special tube-shaped heating tool. Heating is applied until the material reaches the melting stage, followed by inspection of the melting pattern, inserting the pipe end into the socket, and holding it in place until the joint cools. Proper joint design provides sufficient interference fit between the spigot and the socket, so that when the mating surfaces are melted, there is adequate melt to fuse the parts together. Figure 3.36 shows a typical socket fusion joint. This process is mostly used in industrial pipeline construction. Mechanical equipment is available and is commonly used for pipe sizes larger than 40 mm to attain the force required and to assist in alignment (see Fig. 3.37).

Figure 3.36 *Socket joint*

Figure 3.37 Socket welder

The following general steps are included in the socket fusion process:

Joint Preparation

The pipe end and the fitting are prepared for welding by thoroughly cleaning and if necessary, degreasing the surfaces to be joined. A special ring is clamped over the pipe, indicating proper depth of insertion and providing a stopping point when inserting the pipe into
the heating tool and when inserting the pipe into the coupling. It also assists in re-rounding
the pipe.

Heating Phase

Joining surfaces of the pipe and the fitting are brought in contact with the working surfaces
of the heated tool, which has a sufficiently high temperature to rapidly melt the mating
surfaces with minimal heat transfer into deeper layers of the components. For joining PE
components, which is one of the most common applications of socket fusion welding, the
recommended temperature of the heated tool is about 270 °C. Required heating parameters
(temperature and time) depend on the material and the diameter of the components. Guidelines for these parameters are provided by fitting manufacturers and should be followed
closely.

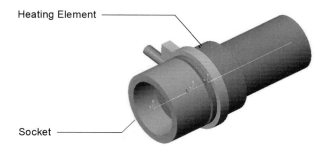

Figure 3.38 Typical socket heater

Change-Over and Fusion Phase

The pipe and the fitting are removed simultaneously from the tool, the melt patterns are inspected for melt layer thickness uniformity, and the pipe is inserted squarely and fully into the fitting, until the fitting contacts the cold ring.

Hold Phase

During the hold phase the joint cools and the assembly should be restrained so the pipe could not come out of the fitting. Cooling times are provided by fitting manufacturers.

3.6.2.3   Saddle/Sidewall Fusion

This technique, illustrated in Fig. 3.39, was developed to join a saddle to the sidewall of the pipe. It consists of simultaneously heating both, the external surface of the pipe and the matching surface of the "saddle"-type fitting with concave and convex shaped heating tool until both surfaces develop an appropriate melt layer. This is accomplished by using a specialized saddle fusion machine that provides the operator with proper alignment and force control.

Figure 3.39 Side wall fusion weld

Side fusion machines are relatively simple devices; however, their design enables the fusion force to be applied through the centerline of the pipe, promoting consistent fusion pressure. A typical machine consists of a sidewall applicator assembly, a heating tool with adapters, fitting holders, and a bottom support. Machines can be operated in horizontal and vertical positions. The Plastic Pipe Institute recommends that sidewall-type joints be made only with a mechanical assist tool unless hand fusion is expressly allowed by the pipe and/ or fitting manufacturer.

The following are typical steps create a saddle fusion joint:

Loading of Parts

The fusion machine is mounted on the pipe using appropriate tooling to round the pipe if necessary.

Preparation of Joint

The joint area on the pipe is thoroughly cleaned, and the pipe surface and the fitting saddle contour are prepared by roughing them with coarse utility cloth or scraping with a special-ized tool to expose fresh material. The cleaning of the mating surfaces assumes special importance due to the fact that the outer surface of the melt in the saddle joint is not displaced from the joint area and all contaminations that are not removed during the cleaning will remain in the joint reducing its strength. Improper preparation of the mating surfaces is one of the major causes of saddle joint failures.

Tool Placement

When proper saddle-fitting adapters are installed in the machine, the fitting is positioned on the pipe and placed into the adaptor. A precise fit between the fitting and the pipe is assured by inspecting the fitting alignment while applying pressure on the pipe and moving the fitting away from the pipe and back to the pipe surface.

Heating Phase

The heating plate is selected for the particular pipe and fitting adaptors and it is positioned between the main pipe and the fitting surfaces. A matching pressure is applied until a consistent melt bead is seen around the joint. At this time, the pressure is reduced to a minimum and the heating cycle is continued in accordance with the fitting manufacturer's guidelines.

Weld and Hold Phase

After the recommended heating requirements have been met, the movable clamp is raised and the heater is removed. Melted surfaces of the fitting and the pipe are brought together with a prescribed fusion force. The pressure is maintained until the joint is completely cooled.

3.7   Material Weldability

3.7.1   Material Influences

Basically, all thermoplastics and all thermoplastic elastomers can be joined by hot plate welding, if their melting range lies below their decomposition temperature. Information about the plastics used most frequently is available (usually from the manufacturers of pipe material) and includes weld strength values to be expected from the various plastics. In case there is no practical experience available, it is advisable to make welding trials with the selected material prior to designing the molding. Such trials can be made with standard-ized specimens. The obtained weld strength values can be considered in the design of the moldings. Good welding results can only be achieved with joining components which have been manufactured in accordance with the processing guidelines for the respective material. The components must be at least at room temperature before welding. Impurities must generally be kept away from the joining surfaces. If necessary, extra processing or cleaning prior to welding can be performed.

3.7.2   Welding of Different Thermoplastics

Hot plate welding can be used to join some dissimilar thermoplastics. Usually, semi-crystal-line plastics can be joined only with compatible semi-crystalline plastics, and amorphous plastics can be joined only with compatible amorphous plastics. It is recommended to make a series of trial welds to measure the weld strength. A similar approach should be used for examining the weld quality of blends.

If the thermoplastic moldings that are to be joined are made of plastics with the same melting point and the same melt viscosity, conventional hot plate welding or – depending on the type of plastic – high-temperature welding can be used. If the thermoplastics to be joined have different melting points or different viscosities, it is recommended to use dual hot plates as described in Section 3.2.5. A typical application example is automotive rear lights: the housing is made of ABS and the lenses are made of PMMA or PC. If the joining parts are made of the same type of plastic, there are no problems. Common material combi-nations include but are not limited to the following:

- ABS – PMMA
- ABS – PC
- ABS – SAN
- PMMA – PC + ABS
- PC – PC + ABS

These material combinations are joined by conventional hot plate welding as well as by high-temperature and radiant heating welding. If both parts have the same melting point, single hot plates can be used. In case of different melting points, it is recommended to use dual hot plates.

3.7.3   Additives

Materials used for the manufacture of moldings frequently contain additives in order to improve their properties or reduce cost. Additives such as stabilizers, lubricants, processing aids, coloring agents, reinforcing materials (talcum, glass fibers, carbon fibers, etc.) can influence the welding properties. Only the plastic itself, i.e., the matrix material can be joined. Additives can have the effect of a stress concentrator in the joint area and therefore reduce the weld strength. It can be assumed that the weld strength will not exceed the matrix strength and it could be low compared to the reinforced thermoplastic bulk material strength. To maintain the same load transfer through the joint of the parts, it is possible to use enlarged joining surfaces.

3.7.4   Recycled Material

If recycled material is added to the basic material, reduced weld strength must be expected. Reground materials were subjected to one or more thermal cycles, which may have caused a degradation of the material. The welding properties of recycled materials should be determined by experimental trials.

3.7.5   Moisture

Different thermoplastics, mainly amorphous thermoplastics, absorb water from the surrounding air. Table 3.2 shows the water absorption of several amorphous thermoplastics. Depending on the water content, this can lead to bubbles forming in the melt during heating and joining, which reduces the weld strength. Therefore, wherever possible, the parts should be welded quickly after injection molding (e.g., within 24 hours after manufacture). If that is not possible, the parts should be stored in a dry environment with water proof bags filled with desiccant.

Table 3.2 Properties of Selected Amorphous Thermoplastics

Material type	Proportion of coherent phase [%]	Injection molding processing temperature [°C]	Melt viscosity (MVR)		Water absorp-tion with sa-uration [23 °C/50%]
			[ml/10 min]	[°C/kg]	
PS		180 to 280	1.2 to 25	200/5	< 0.1
PMMA		190 to 270	0.8 to 21	230/3.8	0.3 to 0.4
PC		280 to 320	3.5 to 18	300/1.2	0.15
PES		320 to 390	6 to 30	300/21.6	0.2 to 0.7
PSU		320 to 390	30 to 100	300/21.6	0.2 to 0.7
ABS		210 to 280	3 to 40	220/10	0.3 to 0.4
ASA		240 to 280	4 to 24	220/10	0.35
SAN		200 to 270	5 to 22	220/10	0.2 to 0.3
SB		180 to 280	3 to 23	200/5	< 0.1
PC+ABS	PC: 40 to 85	255 to 280	8 to 22	260/5	0.2
PC+ASA	PC: 40 to 85	200 to 300	3 to 45	220/10	0.2 to 0.3
PPE+SB	PPE: 20 to 95	240 to 300	24 to 270	250/21.6	< 0.15
PPE+PA	PA: about 50	270 to 300	10 to 50	275/10	0.8 to 1.2
PC+PBT	PC: 40 to 50	255 to 280	8 to 13	250/5	0.35
ABS+PMMA	ABS: 40 to 70	210 to 280	5	220/10	0.3 to 0.4

3.7.6   Welding of Plastics with Different Melt Viscosities

In practice, it is common for the parts to be made from the same thermoplastic, but they can have different melt viscosity. If the same hot plate temperature was used for both par-s, the melt of the plastic part with the lower melt viscosity would be completely forced out of the joint area during joining, whereas hardly any melt flow would occur in the plastic part with the higher melt viscosity, resulting in very poor welds. In this case, optimum welds can only be achieved with a melt viscosity being the same for both parts. In practice, we use dual hot plates, with each hot plate set to a different temperature. This allows the more viscous part to be heated at a higher temperature so that its viscosity decreases and is closer to the viscosity of the second part. Another solution would be to use different heating times with a single hot plate.

3.8   Troubleshooting

Table 3.3 details selected problems associated with hot plate welding and possible solutions.

Table 3.3 Troubleshooting Guidelines

Problem	Probable Causes	Recommendations
Material sticking to heated tool	Anti-adhesive coating is damaged	Hot plate working surface needs to be re-coated
Burning of plastic	Excessive hot plate temperature	Decrease temperature
Poor weld strength	Excessive pressure during the heating time: all molten material is being displaced while parts are heated	Reduce pressure during the heating phase
	Solidification of melt surface	Reduce change over time
	Different melt flow index of materials in joining parts	Heating parameters (temperature and heating times) for two parts should be adjusted to balance the MFI on joining surfaces before forging. Use lower temperature (if dual hot plate system is used), or shorter time, for material with higher MFI
	Excessive pressure during the forging: all molten material is displaced from the weld zone	Reduce joining pressure, or travel during the forging (if machine with displacement control is used)
Small flash	Inadequate heating	Increase heating temperature and/or time

Table 3.3 *Troubleshooting Guidelines (Continuation)*

Problem	Probable Causes	Recommendations
Too much flash	Excessive heating time	Reduce heating time
	Excessive pressure during the heating time: all molten material is being displaced while parts are heated	Reduce pressure during the heating phase
	Excessive pressure during the forging: all molten material is displaced from the weld zone	Reduce joining pressure, or travel during the joining phase (if machine with displacement control is used)
Non-uniform weld	Poor alignment between the parts and the hot plate	Check both, angular and lateral alignment of the parts against the hot plate. Re-align parts if necessary to ensure parallelism of the hot plate working surfaces and the joining surfaces of the parts, as well as the center position of the hot plate on the parts.
	Poor surface quality on the parts	Check the quality of the joining surfaces before welding.
		Increase matching pressure
		Increase matching time
Smoke	Burning of plastic	Reduce temperature
Stringing	Melt sticking to heated tool	Increase tool travel speed/acceleration
		Check anti-adhesive coating on the hot plate
No melt	No melting	Increase tool temperature
No displacement	No contact between the part and the hot plate	Adjust mechanical stops during heating stage
Porosity in flash or bond line	Polymer degradation	Reduce hot plate temperature
		Reduce heating time
	Moisture out-gassing	Dry parts prior to welding
		Increase clamp pressure

Problem	Probable Causes	Recommendations
Unsymmetrical melt	Poor contact between one of the parts and the hot plate	Adjust mechanical stops during heating stage
	Different melt flow index of materials in joining parts	Heating parameters (temperature and heating times) for two parts should be adjusted to balance the MFI on joining surfaces before forging. Use lower temperature (if dual hot plate system is used), or shorter time, for material with higher MFI
No melt	No melting	Increase tool temperature
No displacement	No contact between the part and the hot plate	Adjust mechanical stops during heating stage

3.9   Acknowledgements

The authors would like to thank the following organizations and companies for providing the materials for portions of this chapter:

Plastics Pipe Institute, Washington, DC, USA
Widos GmbH, Ditzingen-Heimerdingen, Germany
T.D. Williamson, Inc., Tulsa, OK, USA.

3.10   References

Bielomatik "Hot Plate Welding Machines and Production Lines"

DVS Specifications "DVS 2215-1, DVS 2215-2 DVS 2215-3"

4 Hot Gas Welding

Ted Hutton

4.1 Introduction

Hot gas welding is a very popular, low cost, versatile means of joining thermoplastics. During the process, heated gas or air is used to heat and melt a filler (weld) rod into the joint. It is a very flexible process, which makes it well suited for short-runs or prototype welding of small items or for welding of large structures or tanks. Hot gas welding can be performed manually, in speed welding mode, and in automated mode. As shown in Fig. 4.1, in manual operation pressure is applied by pushing the weld rod into the joint area by hand. In speed welding, a tip with a pressure shoe or tongue is used to apply pressure enabling higher welding speeds. Figure 4.2 shows the cross section of a welding tip for speed welding and Fig. 4.3 shows a photograph of a speed welding tip. Automated welding is performed using custom equipment and is designed specifically for each application.

The filler rod should be made from the same material as the parts that are to be welded. Usually, the filler rod has a round cross-section, but it is also available in oval, triangular and rectangular cross sections. Like in metal welding, for large joints, multiple passes are used to fully fill the cavity. In some cases it may be desirable to use a small filler rod for the first pass to ensure full penetration and then use larger weld rods for subsequent passes.

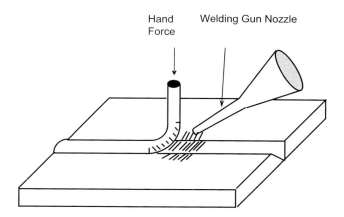

Hand Force   Welding Gun Nozzle

Figure 4.1 Manual hot gas welding

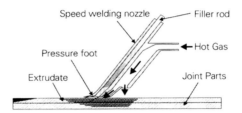

Figure 4.2 Welding tip for speed welding

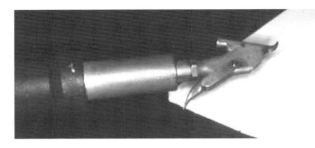

Figure 4.3 Speed welding tip

4.2   Physics of the Process

Hot gas welding, like other welding processes, includes the five basic welding steps: surface preparation, heating, application of pressure, intermolecular diffusion, and cooling (see Chapter 2). Compared to other welding processes, surface preparation is relatively involved for hot gas welding. Depending on joint design, it entails machining the joint area to produce a "V" groove or another configuration and it is usually followed by scraping to remove contaminants. Figure 4.1 shows a "V" groove machined into the plastic plates that are ready for butt welding.

During hot gas welding convection heating of the weld rods and the weld surfaces on the components are used to form the melt. For manual welding, the welding gun nozzle is fanned back and forth allowing heated gas to impinge alternatively on the weld rod and the surfaces of the parts. The rate of heating depends on the gas temperature and velocity. In speed welding, the weld rod is heated by both convection and conduction as it is pushed through a heated channel in the welding tip. In this case, the weld rod acts like heated extrudate that is pressed into the groove. The weld surfaces on the parts are heated by convection as heated gas flows over them.

For manual welding, pressure is applied by the operator pushing down on the weld rod. Depending on the stiffness of the material being welded, the angle that the weld rod makes with the plane of the welding surfaces is changed to apply more or less pressure. In speed welding, the pressure is applied by the pressure foot that is part of the welding tip.

The pressure brings the melted weld rod in intimate contact with the molten surfaces on the parts enabling intermolecular diffusion and chain entanglement. In addition to the intermolecular diffusion, the thermoplastic undergoes significant micro-structural changes as it is cooled. For amorphous polymers, micro-structural changes occur in the form of molecular orientation as the weld area cools. For semicrystalline thermoplastics, the crystallites and spherulites that form depend on the cooling rate. The use of temperatures outside the prescribed range will result in crystalline structures that will be more likely to fail when subjected to stresses. Incorrect temperatures and rapid cooling can result in a significantly smaller crystalline and spherulite structure in the weld zone. The smaller spherulite structure is more likely to fail when exposed to chemicals, solvents, or stress. Therefore, rapid cooling of weld joints is not recommended. The backing material, which is used to support the part during welding, can affect the cooling rate. Welding on materials such as concrete, massive metal plates or other substrates that will rapidly pull the heat from the weld joint should be avoided. Increasing the temperature of the hot gas is not a proper solution to this problem.

It is important to hold the parts firmly in place during welding. Movement of the parts or the joint during the weld process will result in inferior quality welds as the parts are stressed before complete cooling. Simple fixtures can be built to prevent movement of the thermoplastic while allowing easy access to the weld joint.

4.3   Process Description

In hot gas welding a filler or weld rod, which is usually made of the same material as the welded parts (substrates), is heated to its melt point. In addition, the surfaces of the substrates are heated to the melt point of the thermoplastic. Once all faying surfaces are fully heated, the rod is pressed into a prepared joint, promoting squeeze flow and intermolecular diffusion at the molten interfaces. Table 4.1 shows the general steps involved in making a weld. It is important to note that in this table all cycle times depend on the weld length, since it is common for this process to be used for very long weldments (over 10 m). It is also important to note that this process is relatively time consuming, especially when multiple passes are required. Therefore, hot gas welding is not used in mass production, such as assembly of automotive parts. It is well suited, however, for prototype construction as well as for the fabrication of large vessels, pipes, or other structures.

Hot gas welding utilizes a filler material heated to a specified temperature that is pressed into a prepared, preheated joint. As shown in Table 4.2, there are four primary welding parameters for hot gas welding. Other factors, such as joint design, filler rod size, and application of filler rod (such as angle of filler rod introduction relative to substrate) also greatly effect weld quality and are described in later sections. The temperatures utilized in hot gas welding are material-dependent and are very critical to achieving proper fusion. Excessively high temperatures can promote polymer degradation. In contrast, too low a temperature can prevent full melting, squeeze flow, and intermolecular diffusion resulting in a "cold" joint. Tables 4.3 and 4.4 provide a listing of the recommended welding parameters for manual and speed welding respectively. Table 4.5 provides recommended welding parameters for thermoplastics not currently listed in published documents. Another key parameter is travel speed, which depends on the material, gas temperature, and weld rod size. Control of the heat and pressure that are applied to the weld rod along with time through the travel speed are extremely important. If properly performed, hot gas welding can result in weldments that have a tensile strength approaching that of the sheet material strength

Table 4.1 Typical Hot Gas Welding Cycle

Step	Typical time
Machine/prepare joint design	Manual process 3 min/ linear meter weld length
Clean faying surface	Manual process 1 min/ linear meter weld length
Make weld	Manual process 3–5 min/ linear meter weld length
Inspect weld	Manual process 3 min/ linear meter weld length

Table 4.2 Summary of Welding Parameters

Parameter	Influenced by:
Air/gas temperature	Filament temperature, gas flow rate and gas thermal properties
Gas flow rate	Gas pressure and nozzle opening size
Travel speed	Operator
Weld pressure	Operator and tip design

Table 4.3 Recommended Welding Parameters for Manual Hot Gas Welding [1]

Thermoplastic	Welding Temperature[1]	Gas Flow Rate (L/min @ 0.4–0.7 atm)	Force (N) 3 mm	Force (N) 4 mm
Polyethylene HD-PE	300–320 °C	40–60	6–10	15–20
Polypropylene PP-H, PPB, PP-R	305–315 °C	40–60	6–10	15–20
Polyvinylchloride PVC-U	330–350 °C	40–60	8–12	8–12
Polyvinylchloride PVC-C	340–360 °C	40–60	15–20	15–20
Polyvinylidenefluoride PVDF	350–370 °C	40–60	15–20	15–20

1) Measured 5 mm inside weld tip outlet

Table 4.4 Recommended Welding Parameters for Speed Hot Gas Welding [1]

Thermoplastic	Welding Temperature[1]	Gas Flow Rate (L/min @ 0.4–0.7 atm)	Force (N) 3 mm	Force (N) 4 mm
Polyethylene HD-PE	320–340 °C	45–55	10–16	25–35
Polypropylene PP-H, PPB, PP-R	320–340 °C	45–55	10–16	25–35
Polyvinylchloride PVC-U	350–370 °C	45–65	8–12	15–25

Table 4.4 *Recommended Welding Parameters for Speed Hot Gas Welding [1] (Continuation)*

Thermoplastic	Welding Temperature[1]	Gas Flow Rate (L/min @ 0.4–0.7 atm)	Force (N) 3 mm	Force (N) 4 mm
Polyvinylchloride PVC-C	370–390 ºC	45–65	15–20	20–25
Polyvinylidenefluoride PVDF	365–385 ºC	45–65	12–17	25–35

1) Measured 5 mm inside weld tip outlet

Table 4.5 *Recommended Welding Parameters for Speed Hot Gas Welding of Other Thermoplastics*

Thermoplastic	Welding Temperature[1]	Gas Flow Rate (L/min @ 0.4–0.7 atm)	Force (N) 3 mm	Force (N) 4 mm
Ethylenechlorotri-fluoroethylene ECTFE[2]	350–380 ºC	50–60	11–16	TBD
Polytetrafluoroethylene propylene FEP[2][3]	380–390 ºC	50–60	11–16	TBD
Tetrafluoroethylene perfluoromethyl-vinylether MFA[2][3]	395–405 ºC	50–60	11–16	TBD
Perfluoroalkoxy Copolymer PFA[2][3]	400–410 ºC	50–60	11–16	TBD
Polytetrafluoroethylene PTFE/PFA[2][3]	580 ºC	50–60	11–16	TBD

1) Measured 5 mm inside weld tip outlet   3) Special protection required for welders
2) Nitrogen or other inert gas required   TBD To be determined

The recommended hot gas temperatures in Tables 4.3–4.5 represent the typical range to produce relatively strong welds that will result in the maximum service life. Most thermoplastics can be welded above or below this range but the quality of the resulting weld will often be diminished. Accurate measurement of the temperature of the gas should be made using a digital thermocouple with a small probe. Figure 4.4 shows a typical temperature-measuring device. The temperature is usually measured 5 mm inside the welding tip. It is important to properly brace the welding gun and carefully place the probe inside the welding tip during measurement of the temperature to avoid erroneous readings. Whenever the welding tip is changed, the temperature should be measured and adjusted if necessary. It is a good welding Quality Assurance (QA) procedure to measure and record the welding temperature several times during the work day. It is highly recommended that the thermoplastic material being welded be identified on a QA checklist at the beginning of the welding process. The measured welding gas temperature, welder identification, date, time, ambient temperature, type of welding gas and other information necessary to accurately describe the weldment and conditions during the welding should be recorded on the QA checklist. Whenever the operator interrupts the welding process for an extended time, the welding gas temperature should be re-measured and recorded. Measuring and recording the temperature prior to starting the welding process and after stopping the welder can easily detect deviation from the standard welding conditions. Improper welds should be removed and replaced. Failure to utilize proper welding temperatures is the easiest cause of problems in the hot gas welding process to eliminate. Automatic temperature controlled guns provide a consistent temperature but the temperature is measured prior to the weld tip. The size, shape and mass of the welding tip have a tremendous effect on the actual temperature at the weld area. Therefore, it is important to measure the temperature 5 mm inside the weld tip on automatic heat control welding guns as well.

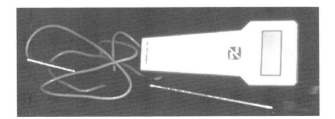

Figure 4.4 *Pyrometer used to measure the hot gas temperature*

The volume of gas going through the welding gun can greatly affect the temperature. Lower air volume can result in increased temperature. For this reason, the standard pressure and volume values listed in Tables 4.3–4.5 should be measured using a pressure gauge and flow meter with a range that will result in accurate pressure and volume measurement.

The amount of force on the filler material is one of the most difficult variables to control. This is dependent on the skill of the operator and the tip design. The ability of the operator to apply the proper force can be determined by having the welder press against a scale to generate the proper pressure. Improper weld pressure will result in a weak joint. Excessive pressure will force the melted polymer out of the weld joint resulting in a "cold" joint. Too little pressure will not allow proper intermolecular diffusion or mixing. Both conditions result in a weak joint.

The proper speed (time) of the weld depends on operator skill, and should be selected based on material, weld rod size, hot gas temperature, joint design, and equipment. Operator skill is usually the most critical variable in assuring the proper travel speed. Operators must develop a "feel" for how a particular thermoplastic welds by evaluating the physical quality of their weld. Typically, too high a speed would result in "cold" welds with the molten weld rod not fully wetting the surfaces of the parts. In this case, the weld rods can usually be peeled of the surface with little or no damage to the parts. Too low a speed usually results in degradation in the weld rod and/or the part's surface, which is evident by charring or gaseous emissions.

4.3.1   Manual Welding

As with standard hot gas welding, this procedure requires control of all of the parameters listed in the welding condition tables, but because of the nature of the process, control is even more critical. This is due to the fact that instead of using a tip design that deflects the heated gas to both the filler rod and sheets simultaneously (as in speed welding, see Section 4.3.2), a single air flow is "fanned" between the two. If performed properly, using a constant back and forth motion of the heat source produces uniform melt in both the sheet and weld rod. This procedure works adequately with rigid thermoplastics such as PVC and CPVC. Other softer polymers such as the polyolefins and fluoropolymers do not possess the rigidity in the weld rod to allow for adequate pressure at the rod/sheet melt interface to produce the desired quality weld. This welding procedure is not frequently utilized due to the availability of new speed welding tips. Manual welding is often erroneously specified in the welding of thermoplastics for high purity applications due to the belief that some metal particles or atoms may transfer from the speed-welding tip to the weld joint. While this may occur to some minute degree, far more transfer of contaminants can occur during the extrusion and manufacturing processes of the components being welded. The resulting reduction of weld quality due to the use of manual welding instead of speed welding is not justified by the immeasurable possible level of contamination resulting from speed welding.

4.3.2   Speed Welding

The use of a speed tip for hot gas welding is the most popular and reliable method for hot gas welding. The filler material is fed into the tip where it is preheated along with the weld joint. This preheating is achieved by re-directing a portion of the heated air/gas so that it passes over the rod before the rod is forced in the joint. Typically, the end of the weld tip has either a curved shape or a roller to help fully force the filler material into the joint (Figs. 4.2 and 4.3). The weld tip is where the pressure is applied during the welding process. Often the skill of the operator defines the quality and consistency of the shape and surface finish of the weldment. It is important to have a proper shape with smooth edges.

Individual weld tips are often modified to optimize the quality and appearance of welds for a particular application. For example, a tip that has been modified for inside corners may not work well with outside corners or butt welds.

The speed tip shown in Fig. 4.3 has a tacking tip (shown folded back) combined with the speed tip. The tacking tip can be moved to the forward position for tack welding. Tack welding uses no filler material and it is used to temporarily hold the sheets or components together until the final weld is made. The tack welding tip is used to melt a small amount of the sheet on both sides of the weld "V" and force that melt into the joint. Tack welding can be continuous, along the entire joint, or discontinuous, only in selected spots along the weld joint. A good continuous tack weld provides an added measure of quality to the process. It allows accurate alignment and holding of the parts. It has also been demonstrated that if the continuous tack weld is tested to be pinhole free, vessels built to contain liquids have a lower failure rate. Proper preparation of the weld joint in the form of scraping prior to welding is necessary to provide a strong finished weldment. Tacking tips for hot gas welding equipment are also available as separate tips that are not incorporated into the speed welding tip.

4.3.3   Welder Consistency

Hot gas welding is highly operator dependent. Therefore, it is essential that the welder have adequate experience with the thermoplastic material being welded and the welding procedure. Control of the movement of the welding gun is necessary for the production of a quality weld. Most thermoplastics will show slight changes in surface appearance compared to the parent material about 5 to 8 mm on either side of the weld joint. This can be observed as a difference in gloss when viewed from a low angle along the axis of the weld. This heat-affected zone (HAZ) should be uniform in size along the length of the weld and from side to side. If one side is wider than the other, this is an indication that the side with the widest HAZ received more heat than the other side of the weld joint. Any observance of this phenomenon is an indication that the welding procedure was not properly followed or controlled. A weakened area of the weld joint occurs at this point. During the normal restarting of the filler material, a short section of wider HAZ will occur. This should be uniform on both sides of the weld zone and the HAZ will resume a normal width shortly after the restart. This is a normal occurrence and not necessarily an indication of a defect. Figure 4.5 shows the HAZ along a weld that has been outlined and darkened with ink to show the HAZ.

Currently, there are no reliable non-destructive tests for the evaluation of thermoplastic welds (see Chapter 15 for more details on testing). The appearance of the weld combined with destructive testing of weld samples is the principle method of assuring welder consistency and quality. Table 4.6 shows a few of the weld defects that are observed in hot gas welds [2]. The American Welding Society (AWS) describes three classes of welds based on the level of requirements for reliability and/or intended service [2]. Class I has a high level of requirement for reliability and/or intended service. Similarity, classes II and III and have medium and low level requirements for reliability and/or intended service respectively. Once the class of requirement for a weld is determined, the AWS document describes various limits for defects. Similar documents exist within other organizations.

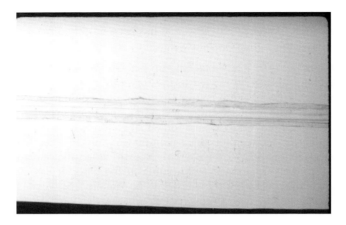

Figure 4.5 Heat affected zone for a hot gas weld

Table 4.6 Imperfections in Hot Gas Welds [2]

External state of joint	Description	Evaluation groups
Cracks	Isolated cracks or groups of cracks with and without branching, running length wise or crosswise to weld. They can lie in weld, base material or in heat affected zone.	Not permissible
Welding flash notches $K > 0$	Market deepening, which can be seen and felt, along or between individual welding rods, caused by, for example faults on die, or poor welding rod guidance.	Not permissible: Group I Locally permissible: Group II if $K > 0$ Group III if $K > 0$
Root not welded through Δs	Notch caused by incomplete weld filling at root, due to, for example: generating angle too small, during edge preparation, root gap to small, root rod too thick, or welding force too small.	Not Permissible: Group I and Group II Permissible: Group III: $\Delta s \leq 0.1\ s$
Weld root too high Δs	Caused by, for example: root gap too big or compressed stringer bead	Permissible Group I: $\Delta s \leq 0.15\ s$ but max. 2 mm Group II: $\Delta s \leq 0.2\ s$ but max. 3 mm Group III: $\Delta s \leq 0.25\ s$ but max. 4 mm

In order to determine an operator's proficiency level and consistency, regular tests of weld samples should be performed. As long as an operator is regularly welding a specific thermoplastic with a specified procedure, testing is recommended on an annual basis. If the operator does not weld the specific thermoplastic using the specific method for six months, retesting should be performed. The German Welding Society (DVS) provides the requirements for qualification [1]. The short-time tensile test and technological bend test are adequate to evaluate a welder. Both tests should be performed according to the DVS procedures and the results must meet or exceed the requirements specified by DVS. Testing of samples prior to a sixteen-hour normalization period after welding will produce erroneous results. Chapter 15 contains more information on testing.

4.4   Equipment

There is a wide range of hot gas welding equipment. Relatively low cost systems have minimal control while more sophisticated machines are nearly totally automated and control all parameters. However, there are several common key components to all systems. as described below.

4.4.1   Hot Gas Gun

The hot gas gun usually includes an electrical heating element, which heats the gas that passes through it. The selection of the welding gun is often made by what is available in the shop or what type of gun is the least expensive. Sometimes the selection gets slightly more sophisticated and the availability of replacement parts or service is considered. The selection of a gun is seldom made on the most important aspects such as ability to maintain temperature, safety, ease of handling, and reliability. Aside from the safety of the equipment, the ability to control the temperature is probably the single most important consideration in the selection of a weld gun. A high quality variable power supply is necessary to control the amount of power to the heating element. It is assumed that the proper size heating element has been selected.

4.4.2   Air/Gas Source

Depending on material being welded, the heated gas can simply be atmospheric air or it can be inert, typically pure nitrogen. The selection of the gas is based on the thermoplastic's tendency to oxidize or degrade during welding. The hot gas welding process does not require high-pressure gas. The use of high-pressure compressors results in a significant increase in probability of contamination of a weld with oil or water in the gas stream. Low-pressure blowers provide adequate air volume and pressure. They are less expensive to operate and maintain and they do not require complex dryers and separators. Evaluation of the air supply can be performed by permitting the air to flow through a clean white cloth for several minutes. Any particles, oil or moisture present in the cloth indicate an inadequate air supply. This must be corrected prior to welding.

To control the gas flow, regulation of the pressure and/or flow rate is required. A typical system, like the one shown in Fig. 4.6 includes some form of pressure regulation for the gas to ensure a low pressure. When using pressure regulation, a flow meter should be added to the system. By recording the volume of air along with the pressure and power settings, it is easier to set up the same conditions the next time the equipment is used. The flow meter will also indicate any changes in the volume flow rate, which will result in a temperature change of the gas. It is important to note that if an inert gas is being used to weld the plastic, it should be used during the setting of the temperature of the welding gun. The difference in heat capacity of gasses will cause the temperature to vary. If a gun is set up for air and the welding is done with inert gas, the temperature will change when the gas source is changed.

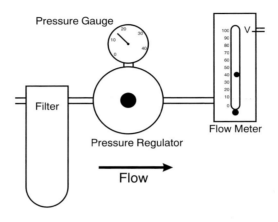

Figure 4.6 *Typical gas pressure and flow regulation system*

4.4.3   Temperature Control

The type of hot gas temperature control used varies greatly depending on the other system components. In some cases, temperature control is simply a current or voltage control of the heating element. Unfortunately, this is not sufficient to fully control the gas temperature because the air-flow rate and the type of gas will also affect its temperature. In more sophisticated systems, an internal temperature measurement device, such as a thermo-couple, measures the actual gas temperature. The signal is then used in a closed-loop circuit to maintain the temperature at the selected value. Typically this control system maintains the set temperature that was measured 5 mm inside the tip when the input of voltage or current, gas type, and gas pressure or flow rate change within normal variations. Most such systems will shut down the heating element if conditions exceed normal varia-tions. This will minimize equipment failures resulting from excessive overheating.

4.4.4   Welding Tips

The selection of the type of tip is critical for the quality of the weld. Some of the various types of tips available are: tacking tip, round nozzle, high-speed tip, tacking/high-speed combination, and high-speed roller tips. Figure 4.7 shows examples of hot gas welding tips. The selection of the type of tip should be made taking into consideration the type of plastic being welded, the thickness of the plastic, the geometry of the part being welded, and the welder's proficiency. There is no one type of tip for a job, in fact, the geometry often dictates the use of several types of tips.

The selection of weld tips for manual or high speed welding is determined by the size and shape of the filler material. The size of the nozzle for manual welding will vary from 3 to 5 mm depending on the size of filler rod and the joint. The welder must select the nozzle that provides adequate heat to the filler rod and parent material. This is determined by the welder's experience and verified by testing of sample joints.

Round Tip

High Speed Tip

Combination
Round/Tacking Tip

Combination High
Speed/Tacking Tip

Tacking Tip

Profile High Speed
Tip for Triangular Rod

Figure 4.7 Type of welding tips for hot gas welding (Courtesy Wegener North America Inc.)

The selection of the speed tip is determined mostly by the size and shape of the filler material. Filler rods for structural joints that must meet specific requirements must be round. The size will be either 3 mm or 4 mm in diameter. A root weld of 3 mm rod is required for all structural quality welds. Depending on whether the joint is a single V or double V, the subsequent filler rod will be either 3 mm or 4 mm in diameter.

Another feature that may be helpful in some cases is the addition of a small light on the welding gun. The light provides the best possible illumination of the weld because it is a consistent source of light and it eliminates shadows. Some disadvantages include the additional size and weight that it adds to the gun, the inconvenience of the light in tight places and the problems encountered in finding a suitable light for your welding gun. Some welding equipment suppliers have lights that can be added to their guns. The light is especially useful in welding bonded liners in large tanks and as a training aide for welding classes.

4.5   Joint Design

Structural hot gas welded joints require very specific joint designs. Figure 4.8 shows a single "V" joint design and Fig. 4.9 shows a double "V" joint design. Figures 4.10, 4.11, and 4.12 show T joints, a corner joint, and a lap fillet joint respectively. In preparation of the joint, in addition to the proper beveling of the joint, it is necessary to physically scrape the joint and weld rod just prior to welding. This must be done prior to each weld pass. The physical scraping removes contaminants that are on the surface of the filler material resulting from the manufacturing, handling, and storage of the filler. Scraping of the joint removes impurities caused by the preparation of the joint, handling, and previous weld passes. Solvent wiping spreads any contamination along the surface rather than removing the contamination. Some solvents will penetrate the thermoplastic joint. When heated to welding temperature, the solvent can alter the microstructure of the thermoplastic joint, causing premature failure.

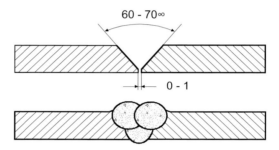

Figure 4.8 Example of single "V" joint designs

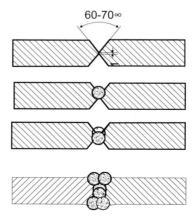

Figure 4.9 Examples of double "V" joint designs

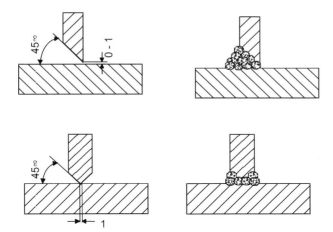

Figure 4.10 *Examples of T joints*

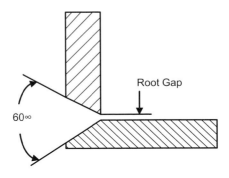

Figure 4.11 *Corner joint*

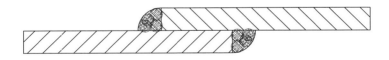

Figure 4.12 *Lap fillet joint*

4.6   Applications

Because hot gas welding is a relatively slow manual process, it is not used in mass production such as parts for consumer products. However, because it can easily be tailored to work with large parts (over 3 m is size) with nearly any geometry, it is used to fabricate pipelines, pond liners, and a wide variety of vessels.

Hot gas welding is found in a variety of applications ranging from chemical plants, semiconductor chip manufacturing equipment, transport vessels, and large storage vessels. Interior support sections for scrubbers are one popular use for hot gas welded thermoplastics (see Fig. 4.13). Wet benches for computer chip manufacturing and other equipment utilized in clean-rooms, where metallic construction is not suitable require considerable hot gas weldments. Figure 4.14 shows a typical wet bench for computer chip manufacturing.

Figure 4.13 Support section for interior scrubbers

Figure 4.14 Wet bench for computer chip manufacturing

Hot gas welding is widely utilized in the fabrication of large vessels that cannot be produced using other welding procedures. Vessel sizes in excess of 325 cubic meters are not uncommon. Most large vessels are metal structures with thermoplastic liners bonded to the metal where the seams are welded using the hot gas welding process. Another very successful fabrication technique is the combination of the welded thermoplastic supported with a thermoset glass reinforced shell. Examples of thermoplastic weldments utilizing the hot gas welding procedure for lining of the interior of railroad tank cars and trailer tanks are shown in Figs 4.15 and 4.16. Large dual laminate lined process and storage vessels are also typical applications.

Figure 4.15 Lined railroad tank car

Figure 4.16 Lined trailer tank

The success of these applications is due to the proper control and evaluation of the welding process and welder's skill. Adequate test procedures and standards exist to evaluate thermoplastic welds. Improvements to the hot gas welding equipment over the past several years have greatly increased the reliability of the hot gas welding process when proper procedures are followed.

Other applications that are typically welded with hot gas welding are pond liners and gas distribution pipelines. The main advantage of using hot gas welding for these applications is ease of operation in remote locations. In addition, the process is easily configured to a wide range of joint designs and application configurations. It should be noted however, that hot gas welding of thermoplastics is usually limited to sheet type applications. The only requirement is that the operator can perform the weld according to procedures or standards. The limitation to the process in these applications is speed (cycle time) and the fact the weld quality is mostly defined by operator skill.

4.7   References

1. DVS Verlag GmbH (German Welding Society). http://www.dvs-verlag.ed/en/index.htm

2. American Welding Society (AWS) G1.10M:2001, Guide for the Evaluation of Hot Gas, Hot Gas Extrusion, and Heated Butt Thermoplastic Welds.

5   Extrusion Welding

Avraham Benatar

5.1   Introduction

Extrusion welding was invented during the early sixties with a number of different variations that were subsequently developed. Its main characteristic is that the weld (or filler)
rod consists of a homogeneous and completely plasticized or molten strand (extrudate) that
is applied directly into the welding joint without being cooled. The plasticized strand is
produced with an extruder or by means of an extruder-like mixing chamber. Prior to introduction of the extrudate, the faying surfaces in the base material are heated to the welding
temperature by hot air. In some cases, radiant heat from halogen lamps may be used instead
of hot air. Through this heating, the weld faces are transformed into the molten state. The
pressure exerted on the melt causes the material from the weld rod and base plates to be
pressed together and to produce a firm joint after cooling [1].

There are a few variations of the extrusion welding method. Although they differ with
regard to technique, equipment, and application, they possess the following common characteristics:

- The filler or weld rod emerges from an extruder unit in the form of a strand.
- The filler rod is homogeneous and completely molten or plasticized.
- The fusion surfaces of the parts being welded are in a molten state from the surface
  down to a certain depth.
- The welding procedure is performed under pressure, thus permitting a certain degree of
  flow to take place.

Extrusion welding is typically done in one pass unless the parts are very thick. Therefore,
extrusion welding typically takes much less time compared to hot gas welding of thick
sections. Table 5.1 compares the times required to produce a 1 m long single-V butt weld in
high density polyethylene (HDPE) using extrusion and hot gas welding. Extrusion welding
takes about 1/6 of the time needed with hot gas speed welding. Therefore, labor costs for
extrusion welding can be substantially lower, making it a more economical process for
some applications [2].

Table 5.1 *Comparison of Production Times for Extrusion and Hot Gas Speed Welding of a 1 m Long*
Single-V Butt Weld of a 25 mm Thick HDPE Sheet [8]

	Extrusion welding [min]	Hot gas speed welding [min]
Specimen preparation (cutting the groove)	12	12
Preparation for welding (clamping and tacking)	15	10
Welding	15 (1 pass, 2 welders)	264 (32 passes with 3 mm weld rod, 1 welder)
Trimming	10	5
Total production time	52	291

5.2   Process Physics

Extrusion welding is very similar to hot gas speed welding, except that an extruder is used to produce the weld or filler material. Extrusion welding includes the five basic welding steps: surface preparation, heating, application of pressure, intermolecular diffusion, and cooling (see Chapter 2). Surface preparation for extrusion welding entails machining the joint area to produce a "V" groove or another configuration and it is usually followed by scraping to remove contaminants. Figure 5.1 shows two plastic sheets with "V" grooves that are being welded.

During extrusion welding the weld rod and the base material are heated in different ways. The weld rod or extrudate is heated by conduction as it passes through the heated extrusion chamber and through heated tubes leading to the welding shoe. While in the extrusion chamber, the extrudate undergoes viscous heating from the shearing action of the screw. The base material is usually heated by convection as hot air flows over the surface. Increasing the air flow rate or increasing the air temperature increases the surface temperature of the base material and produces thicker melt layers [3].

For hand held extrusion welders and movable welding heads, pressure is applied by the operator pushing down on the welding shoe. The pressure brings the molten filler material or extrudate in intimate contact with the molten surfaces of the parts, enabling intermolecular diffusion and chain entanglement. In addition to the intermolecular diffusion, the thermoplastic material undergoes significant micro-structural changes as it is cooled. For amorphous polymers, micro-structural changes occur in the form of molecular orientation as the weld area cools. For semicrystalline thermoplastics, the crystallites and spherulites that form depend on the cooling rate. The use of temperatures outside the recommended range will result in the formation of crystalline structures that will be more likely to promote failure when subjected to stresses. Incorrect temperatures and rapid cooling can result in a significantly smaller crystalline and spherulite structure in the weld zone. The smaller spherulite structure is more likely to promote failure when exposed to chemicals, solvents, or stress.

Figure 5.1 Extrusion butt welding of plastic sheets (Courtesy Wegener North America Inc.)

5.3   **Process Description**

The process of extrusion welding involves the following basic steps:

- Specimen preparation and cleaning,
- Clamping and tack welding,
- Equipment startup,
- Welding,
- Trimming or finishing.

Specimen preparation usually involves machining a groove with an included angle between 45° and 90°, depending on the material thickness and the joint design (see Section 5.5). Electric hand planes, milling cutters, or other machining tools can be used for this purpose. Following machining, the weld area may need to be cleaned to remove any oxidized surfaces or remaining lubricants, which is usually done by scraping. When doing repair work it is important to completely remove sections that have been attacked by chemicals prior to preparation of the joint area. The specimens are then clamped or tack-welded so that they remain firmly in place during welding. It is important to maintain appropriate root gaps as determined by the joint design in order to get full penetration welds.

Equipment startup involves turning all heating units on and allowing the equipment to reach equilibrium temperature conditions. For the heated air system, it is important to allow air to flow through the heaters prior to turning them on. Sufficient time should be permitted to stabilize the air flow rate and temperature. The temperature of the hot air is measured by inserting the tip of the temperature probe 5 mm inside the nozzle. For the weld or filler rod, turning the heaters on allows any remaining materials in the extruder and the tubes to be remelted prior to purging (if changing materials) or welding. To avoid damage to the equipment during this phase, particular care should be taken to ensure that no places (e.g., at the nozzle, at the welding head, at the extrudate outlet) remain unheated. If necessary, these must be heated with ancillary devices or auxiliary blowers. For safety reasons, some extrusion welders are equipped with an interlock device which prevents the drive motor from being switched on too early.

During operation of the extruder, heat is produced by electric heaters on the extruder barrel and by the shearing action of the screw. Therefore, the extrudate will be at a higher temperature than the setting of the electric heaters. It may be necessary to purge some material to allow the extrudate to reach a stable temperature. The extrudate temperature is checked by placing the temperature probe at the outlet from the welding shoe. Fixed thermometers do not supply exact temperature measurements. However, they can serve to identify slow changes in temperature not immediately discerned by the welder (e.g., temperature drop due to breakdown in the heating unit during welding). Filler material that may have over-heated during startup may be thermally degraded and it should be discarded.

The weld or filler material, whether in the form of pellets or coiled strand (wire) must be compatible with the base material and it must be kept dry and clean. Coiled strands or wires must be of good quality and free of voids. For pellets, electrostatic charge build-up may result in open containers attracting dust or other particulates. Therefore, it is preferred to keep pellets in enclosed containers.

Preheating of the base material is usually done with hot air using a conventional hot gas gun. The amount of hot air required depends on the shape of the nozzle. All-purpose nozzles with round or oval cross-sections require larger quantities of air than special nozzles adapted to the individual weld shape. In some cases they are designed so that they project into the weld cross-section [4]. It has been found that having a uniform melt layer thickness helps assure high quality welds [5]. The thickness and uniformity of the melt layer can be checked by inserting a thin metal wire into the melt directly in front of the welding shoe or by scratching the melt with a pin or small screwdriver. Table 5.2 lists the typical extrusion welding conditions including hot air temperatures for HDPE and polypropylene (PP). When welding films, hot air may cause shrinkage of the films as well as waviness. Therefore, preheating of films is sometimes done using halogen lamps with a parabolic reflector.

Table 5.2 *Typical Welding Parameters for Extrusion Welding of HDPE and PP [6]*

Extrusion welding of HDPE and PP	
Material thickness[1] (for single-pass welding) [mm]	0–15
Material temperature[2] [°C]	200–300
Hot air temperature[3] [°C]	250–300
Air flow rate [l/min]	≥ 300
Extrusion rate [kg/hr]	0–2

1) if thicker material is to be welded, multiple passes can be used
2) measured with a penetration thermometer at the extrudate outlet
3) measured at the nozzle outlet

Extrusion welds are normally produced in a single run. Double-V butt welds, or other forms of welds with back-up, require two passes, typically one on the front followed by a second on the back. The ability to use a single pass helps reduce production time and it usually increases consistency and repeatability of the weld. If the welding gap cannot be filled in one pass because the part is too thick, then the output of the welding extruder is too low or the necessary welding pressure too high. In those cases, multiple passes may be used. However, after each pass, the flash needs to be removed to avoid producing notches in the weld.

Subsequent machining of welds is generally not necessary as long as the resulting weld is uniformly smooth with notch-free edge zones. When the welding shoe has become misshapen through wear, or the geometry of the parts to be welded is complex, extrudate may leak out at the sides of the weld shoe (see Fig. 5.2). This causes flash to form, which has no bond with the base material and must be removed or it will create a stress notch and result in weaker joints. A scraper is used to remove the excess material from the edges while the polymer is still moderately warm.

Quality assurance for the weld requires careful attention to the design, welding parameter settings, welder knowledge and experience, and environment. The design of the parts and

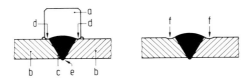

Figure 5.2 *Finishing the weld edges (a – welding shoe, b – base material, c – weld or filler material, d – unwelded area, e – root, f – worked edge zone) [8]*

joints must be specifically adapted to the operating sequence of the extrusion welding process. Specifically, the design should include easily accessible joints allowing the welder to reach all welding seams. The product to be welded and the filler material must be suitable for the extrusion welding process and they must be either identical materials or compatible materials. The welding equipment must be maintained in working condition to fully perform all the functions incorporated into it by the manufacturer. The welders must be experienced in extrusion welding and familiar with the welding apparatus. The welding parameters must be carefully chosen with the help of existing standards and recommendations to insure optimum welding conditions. When welding new materials, simple optimization welds and tests can be performed to find suitable welding parameters. Welding should, if possible, be carried out in an enclosed facility. If circumstances do not permit this, care must be taken to ensure that the working area is protected against adverse weather conditions and that the parts remain clean throughout the welding process.

5.4   Equipment

There are a few variations of extrusion welding that are utilized for different applications. While each variant is slightly different, all extrusion welding systems include the following basic equipment:

- Extruder to produce the weld rod or filler material,
- Hot gas (or radiant heat) source for preheating the base material,
- Welding shoe to press the extrudate into the joint area.

Below is a description of each of the variants of extrusion welding along with the applicable equipment.

5.4.1   Stationary Extruder with Movable Welding Head

The stationary extruder with a movable welding head is used for continuous welding (see Fig. 5.3). The apparatus consists of a pellet-fed single-screw extruder with an adjustable barrel heating system and continuously variable screw speed. The welding speed is determined by the melt flow rate, which depends on the screw rotation speed. A flexible and heated tube that is about 2 m long is used to transport the molten extrudate from the extruder to the welding head. The tube is thermally and electrically insulated to protect the welder. On start-up, the tube is heated to melt the solidified strand in the tube prior to welding. During operation, the continuous heating of the tube prevents the extrudate from cooling and re-solidfying.

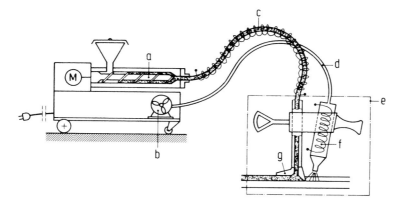

Figure 5.3 *Stationary extruder with movable welding head (a – extruder, b – air supply unit, c – heated flexible tube, d – air tube, e – welding head, f – hot-gas gun, and g – welding shoe) [8]*

An air supply is connected by a second tube to the welding head. The welding head consists of an assembly of the extrudate tube, air heater, welding shoe, and handles for the operator to move it and apply pressure. During the welding operation, the extrudate is pressed into the welding groove with a welding shoe made from polytetrafluoroethylene (PTFE). The air heater is usually a conventional hot gas welding appliance that is installed in front of the welding shoe. It preheats the base material prior to the application of the extrudate to the joint. The air supply for the air heater is either a blower or a compressed air supply. The welding parameters – melt temperature, melt throughput, hot air flow rate, and hot air temperature – are adjustable, thus enabling the welding operation to be optimized for high quality reproducible welds.

5.4.2   Manual or Hand-Held Extrusion Welder

The manual or hand-held extrusion welder is small and lightweight and it is designed for portability (see Figs. 5.4 and 5.5). It includes a mini-extruder driven by a motor similar to those used in small machine tools or drills. The extruder is fed either pellets or coiled welding rod. A coiled welding rod is preferred because it is not affected by the orientation of the welder allowing welding vertically or upside down. The base material is preheated by a conventional hot-gas welding apparatus that receives its air supply from a compressed air system or air blower. Hand-held extrusion welders with an integral air supply are also available. The welding shoe is mounted directly at the output orifice of the extruder. Hand-held welders vary in size and extrudate output with a weight ranging from about 3 kg to 12 kg. Hand-held extrusion welders enjoy the greatest use because they are compact and highly flexible enabling usage in tight areas for a wide range of applications. Operator experience and proficiency are important for consistent production of high quality welds. Welder certification is available in some parts of the world and is under consideration in the US and Europe.

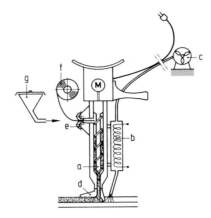

Figure 5.4 Manual or hand held extrusion welder (a – small extruder, b – hot gas torch, c – air supply unit, d – welding shoe, e – rod feed, f – reel, g – pellet hopper) [8]

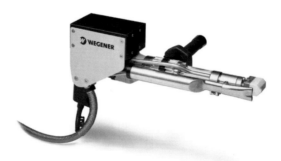

Figure 5.5 Wegener hand-held Alpha extrusion welder (Courtesy Wegener North America Inc.)

5.4.3   Film Welding System

A highly automated mobile welding system is used for continuous welding of films. It was developed specifically for welding of HDPE sheets and films. The apparatus consists of a motor-driven trolley, which carries a single-screw extruder for melt delivery (see Fig. 5.6). The films are welded using a lap joint, where the joint surfaces have been cleaned by a wire brush. The nozzle of an air heater, which in inserted into the overlap, heats the joint surfaces to the weld temperature. A downstream die, which is fed from the extruder, extrudes the filler rod between the melted zones of the films. Preheating of the films extends beyond the width of the extrudate, in order to obtain a good bond with the base material at the edges of the filler rod. Heat input to the films should be minimized to avoid shrinkage or waviness at the edges. This is done by exposing the material for a short time to a relatively high air temperature, which also results in high welding speeds. The necessary welding pressure is applied by a roller system, which is designed to accommodate any irregularities in the surface of the substrate.

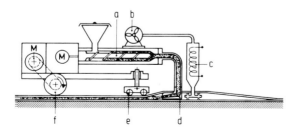

Figure 5.6 *Film extrusion welding system (a – extruder, b – air supply unit, c – hot-gas torch, d – die, e – pressure rolls, f – drive) [8]*

5.4.4   Welder with Manual Transfer of Filler Material

An extrusion welder with manual transfer of the filler material is used for intermittent welding of sections. The system includes a high-output extruder, PTFE tubes to hold the extrudate or filler material, conventional hot-air blowers, and pressure rollers and pressure applicators (see Fig. 5.7). The welding operation itself is carried out intermittently as follows:

The surface of the prepared gap is heated by hot air over a length of approximately 500 mm resulting in melting of the surfaces. Heating of the material is carried out by moving the hot-air blower to and fro in a fanning motion. Shortly before the heating operation is finished, a strand of molten filler rod is conveyed from the extruder into a flexible PTFE tube, which prevents the filler material from cooling before welding commences. During welding, the filler material is allowed to flow from the tube into the weld gap, where slight pressure is applied to it by means of a roller mounted on the tube or by a pressure applicator. Using the pressure applicator, and starting at one end, the filler material is then pressed firmly into the groove in the direction of welding and the welded seam is smoothed and shaped until the filler material has cooled and resolidified. This rolling motion on the

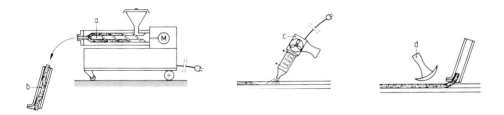

Figure 5.7 *Extrusion welder with manual transfer of filler material (a – extruder, b – PTFE tube for conveying filler material, c – hot-gas torch, d – pressure applicator) [8]*

filler rod ensures that the necessary pressure is applied and that any entrapped air is expelled. This pressure also prevents the formation of shrinkage cavities. The filler material at the end of the weld is tapered to ensure good overlap with the next weld section.

5.4.5   Moveable Welder with Melt Chamber

This welding system includes a melt chamber that has neither barrel nor screw. It consists of a mobile supply unit and a welding head, which are joined by a cable and a conveying tube for the welding rod (see Fig. 5.8). The welding rod is unwound from the reel and is fed to the welding head, where it is pushed into a heated melt/mixing chamber by means of counter-rotating drive rollers. The speed of the drive rollers can be adjusted to change the melt flow rate. After melting in the chamber, the filler material flows to the weld shoe. The solid weld rod, which is fed into the melt chamber, acts as a piston that pushes the melt out at the weld shoe. The base material is usually heated by a halogen lamp with parabolic reflector. This welding system can also be equipped with a conventional hot-air blower. The welding speed can be varied by adjusting the rate at which the rod is conveyed into the melt chamber. For some applications, the weld system has also been partly automated with the welding head being mounted on a trolley that is moved at a constant velocity. This weld system is used to weld flat sheets, rectangular tanks, and sockets on insulating sleeves for district heating lines.

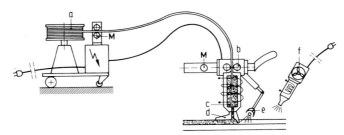

Figure 5.8 *Extrusion welding unit with melt chamber (a – welding rod on reel, b – rod feed, c – melt chamber, d – welding shoe, e – IR lamp or hot-gas torch) [8]*

5.4.6   Welding Shoes

Welding shoes are made from PTFE sheet material to provide low adhesion surfaces that the melt will not stick to. Their design is adapted to the individual weld form and the different material thicknesses to be welded. They are easily interchangeable and possess two lateral guides which prevent the melt from escaping at the sides. With the welding shoe, the pressure necessary for welding is applied to the joint faces via the filler material. The length of the welding shoes must be sufficient to ensure that the welding pressure is applied long enough to allow sufficient cooling to maintain the formed weld faces and compensate for shrinkage of the filler material. The "nose" at the front of the welding shoe prevents filler material from flowing forward. The pressure of the melt against the "nose" helps push the welding head in the welding direction. Figure 5.9 shows a typical weld shoe design and Table 5.3 gives the minimum recommended dimensions for welding shoes. Before welding commences, the welding shoe must be heated, since a rough, uneven weld surface is obtained if welding is done with a cold shoe. The hot air around the welding head, or the hot extrudate that flows past the welding shoe on a piece of wood or plastic sheet, are suitable means of heating the shoe. The shoe can also be heated separately on a hot plate or with a conventional hot-air blower.

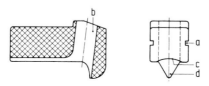

Figure 5.9 *Welding shoe (a – support groove, b – melt feed, c – smoothing track, d – nose) [8]*

Table 5.3 *Minimum Recommended Dimensions for Welding Shoes*

Linear dimensions Pressure-applying L_A = length L_N = length of shoe nose	Thickness a of single-V butt welds [mm]	Lengths [mm]	
		$L_A \geq$	L_N
	up to 15	35	10
	> 15 to 20	45	15
	> 20 to 30	55	20

5.5   Joint Design

Figure 5.10 shows a typical single-V butt weld with labeling of important features. The dashed lines show the preheating zone that is developed from hot gas heating. Also shown are the root and flank of the weld and the overlap. Generally, little or no sideway propulsion of extrudate beyond the overlap occurs and no post welding operations are needed. However, if significant leakage does occur as a result of wear of the welding shoe, then it must be scraped to remove any notches (see Fig. 5.2).

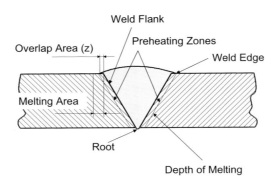

Figure 5.10 *Important features in a single-V butt weld*

Generally, extrusion welding produces one-pass seams. Sheets with difficult accessibility and thick-walled sheets are welded with more passes. Figure 5.11 shows the joint design for a single-V butt joint. To enable full penetration welds, the opening angle for the V groove is 90° for thin plates and 45° for thick plates. A root gap of 2 mm should also be used to insure complete heating and penetration all the way to the root of the weld. While single-pass welding is preferred, for plates that are more than 30 mm thick, multiple passes must be used because the required joining pressure cannot be applied by the operator in a single pass.

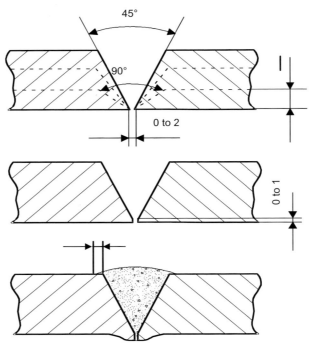

Figure 5.11 *Design of a single-V butt weld*

Figure 5.12 shows the welding sequence for a double-V butt weld. The top part shows the prepared double-V butt weld prior to welding. The second part shows the joint after the upper pass is completed. The third part shows the flipped plates after reworking of the weld root. The bottom part shows the completed double-V butt weld.

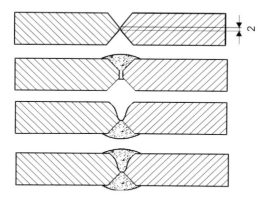

Figure 5.12 Joint preparation and weld sequence for a double-V butt joint

Figure 5.13 shows the weld groove for a single-bevel T joint. Figure 5.14 shows a double-bevel fillet T joint. In both cases it is important to maintain a root gap of about 2 mm to be able to achieve complete penetration. Figure 5.15 shows typical overlap seam welds for films and thin plates; these types of joints are used for geomembranes.

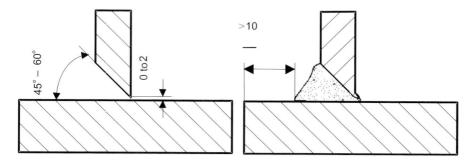

Figure 5.13 Single-bevel T joint

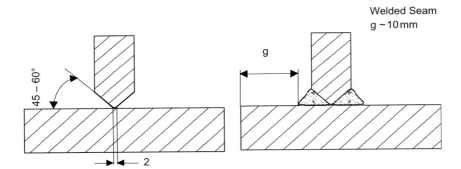

Figure 5.14 *Double-bevel fillet T joint*

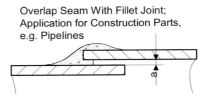

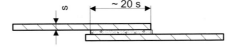

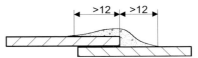

Figure 5.15 *Overlap joints*

5.6   Applications

Extrusion welding is utilized in a wide variety of industries and applications including the following areas of semi-finished product processing:

* In tank construction, where the main task is joining thick-walled parts. This is the case if a tank wall has to be welded to a tank floor, or if flat sheets – which for production reasons cannot be manufactured in the required size – have to be welded together from small sections. Figure 5.16 shows PP scrubbers, which were extrusion welded. Other examples include the welding-on of pipe connections or loose flanges and the attachment of struts or reinforcements.

Figure 5.16 *Scrubbers made of PP for the plating industry (Courtesy Hankwood Line Manufacturing, Inc.)*

* In civil engineering, when heated-tool butt welding is not possible or the fusion faces at the weld areas are not flat. Examples include fabrication of structures, construction of shafts and collecting pits for effluent, welding of additional junctions in existing pipelines, and joining of pipe sections. Figure 5.17 shows extrusion welded HDPE manholes and Figure 5.18 shows an extrusion welded pickling tank for hot dip galvanizing made of PP copolymer. With large-sized pipes with diameters exceeding 1.5 m, extrusion welding is the only technique possible, because there are no heated-tool butt-welding devices suitable for the purpose. Extrusion welding has also been used successfully for the lining of timber and concrete pits, the welding of brickwork collars onto pipelines, and in the repair of damaged pipelines.

Figure 5.17 *Extrusion welded HDPE manholes (Courtesy Plastic Fusion Fabricators, Inc.)*

Figure 5.18 *Extrusion welded pickling tank for hot dip galvanizing made of PP copolymer (Courtesy APR Plastic Fabricating)*

- In apparatus engineering, in cases where the geometric shape of the parts prevents the use of the heated-tool butt-welding method and hot-gas welding cannot be used for quality and consistency reasons (e.g. in the production of filter plates, bases, trays, and other types of construction.)

- In pipeline engineering, when heated-tool butt welding is not possible or possible only under very difficult conditions. For example, when producing pipe bends from segments or from T-joints with junctions of various diameters, which have to be welded in at different angles (see Figs. 5.19 and 5.20). Extrusion welding is also indispensable for the production of reducers from pipe segments or sheet material.

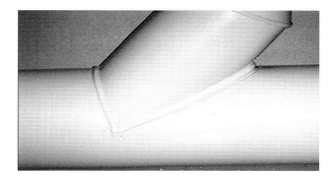

Figure 5.19 *Extrusion welded PP containment lateral in dual wall piping system (Courtesy Asahi America, Inc.)*

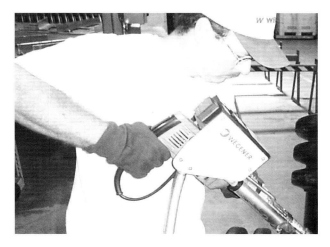

Figure 5.20 *Extrusion welding of cornigated HDPE fitting (Courtesy Advanced Drainage Systems, Inc.)*

• For special tasks, mainly involving the welding of films, webs, or thin sheets. Examples include linings for ponds, linings for protection of buildings against groundwater from outside or leakage of dangerous media from inside. Another important field of application is the lining for waterproofing of large-area refuse dumps to protect sub-surface water [7].

5.7   References

For the convenience of the reader the English titles of all publications in languages other than English are shown in parentheses.

1. Deutscher Verband für Schweißtechnik e.V., Richtlinie DVS 2209, Part 1: Schweißen von thermoplastischen Kunststoffen, Extrusionsschweißen, Verfahren – Merkmale (Welding of Thermoplastic Materials, Extrusion Welding, Processes – Characteristics) (1981)

2. *Gumm, P., Hausdörfer, D., Muth, W.*, Extrusionsschweißen, ein neues Verfahren zum Verbinden dickwandiger Teile aus Hart-Polyethylen (Extrusion Welding, a New Process for Joining Thick-walled Rigid Polyethylene Parts), Kunststoffe **61** (1971), pp. 108/114

3. Michel, P., "An Analysis of the Extrusion Welding Process," *Polymer Engineering and Science* (1989) **Vol. 29**, No. 19, pp. 1376–81

4. *Bemelmann, K.:* Optimierung der Extrusionsschweißnaht durch gezielte Vorwärmung der Fügeflächen und geeignete Ausbildung des Schweißschuhs (Optimization of Extrusion Seam Welding by Selective Prewarming of the Joint Surfaces and Proper Construction of the Welding Shoe). DVS-Berichte, Vol. 84.

5. Gehde, M. and Ehrenstein, G.W., "Structure and Mechanical Properties of Optimized Extrusion Welds," *Polymer Engineering and Science* (1991) **Vol. 31**, No. 7, pp. 495–501

6. Deutscher Verband für Schweißtechnik e.V., DVS 2207-4, Welding of Thermoplastics, Extrusion Welding Panels and Pipes (1993)

7. *Knippschild, F. W., Taprogge, R., Tronow. K.:* Großflächen-Dichtungselement aus Niederdruck-Polyethylen (Large-area Watertight Membranes from Low-density Polyethylene), Kunststoffe im Bau **4** (1977), pp. 154/160

8. John, P. in Heusen, W.

6   Implant Induction (Electromagnetic) Welding

Avraham Benatar

6.1   Introduction

Implant induction welding of thermoplastics is accomplished by inductively heating a gasket that is placed at the weld line. The gasket is usually a composite of the polymer to be welded with conductive metal fibers or ferromagnetic filler. In an alternating magnetic field the filler in the gasket heats resulting in melting of the polymer in the gasket and on the surface of the two parts [1]. The polymer in the gasket must be either the same or compatible with the welded material to enable chain diffusion and entanglement. Then the electromagnetic field is turned off and the parts are allowed to cool under pressure. The gasket becomes a permanent part of the assembly (see Fig. 6.1).

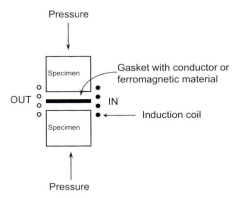

Figure 6.1 Setup for implant induction welding

Implant induction welding of composite materials with or without electrically conductive fibers is also possible [2]. Welding of composites without conductive fibers can be performed by using a gasket that is made of a layer of conductive fibers impregnated with the thermoplastic. Welding of composites with conductive fibers is also possible both, with [2] and without [3, 4] susceptors.

Implant induction welding of thermoplastics and thermoplastic composites is used in a wide range of applications such as automotive, appliance, consumer products, medical, and filtration. Implant induction welding offers a number of important benefits including the following:

- Remote heating of the joint area
- Ability to control heating and cooling rates
- Repeated heating for repair and even disassembly
- Produce hermetic seals and structural joints
- Can join large and small parts in single step or by scan heating
- Can accommodate complex joints with multiple curvatures
- Can join dissimilar materials

Some of the disadvantages of implant induction welding include

- Gasket remains at the joint interface
- Consumable gasket adds to the cost
- Coil design is important
- Maintaining uniform distance between coil and gasket is important for uniform heating
- Impedance matching between system components is important

6.2   Process Physics

Implant induction welding relies on electromagnetic heating of non-magnetic electrical conductors (e.g. aluminum and stainless steel) or ferromagnetic materials (e.g., iron, carbon steel, and ferromagnetic ceramics). When these materials are placed in an alternating electromagnetic field they experience Joule (resistive) heating as a result of induced eddy currents [5, 6]. When electrical conductors are placed in a magnetic field, eddy currents are generated in the conductor so as to negate the field (see Fig. 6.2). From Maxwell's equations it can be shown that for a semi-infinite solid placed in an alternating magnetic field, the eddy current density decreases exponentially along the depth of the solid (see Fig. 6.3) [6]. The current density as a function of depth (z) and time (t) is given by the following relation,

$$J(z,t) = J_s \cdot \exp(-\alpha \cdot z) \cdot \cos(\omega t - \alpha z) \qquad (6.1)$$

Where J is the current density, J_s is the surface current density, α is the reciprocal of the depth of penetration (δ), and ω is the radial frequency of the magnetic field. As shown in

Figure 6.3, there is a certain depth where the current density drops to $e^{-1} = 0.368$ of its surface value. This is referred to as the depth of penetration (δ) or skin depth, which is given by the following relation,

$$\delta = \sqrt{\frac{2\rho}{\omega\mu}} \tag{6.2}$$

Where ρ is the resistivity of the slab and μ is the magnetic permeability. Depth of penetration is important because 86.5% of the losses occur within $z \leq \delta$, and 98.2% of losses occur within $z \leq 2\delta$. This becomes important when considering uniformity of heating. As can be

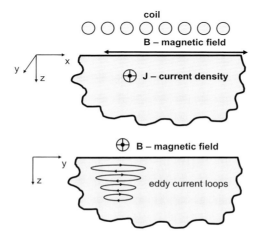

Figure 6.2 *Induced eddy currents in an semi-infinite slab that is placed in a magnetic field*

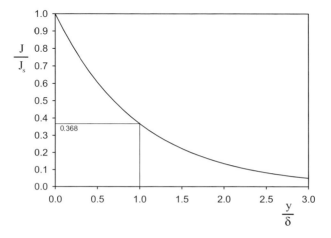

Figure 6.3 *Exponential decay of induced currents in a semi-infinite slab*

deduced from Eq. 6.2, increasing the frequency decreases the depth of penetration. There-fore, at high frequencies the gaskets must be relatively thin and at a uniform distance from the coil to produce uniform heating.

For ferromagnetic materials, hysteresis heating occurs in addition to Joule heating from eddy currents [5, 7]. Figure 6.4 shows the relationship between the magnetic field (B) in a ferromagnetic material and the magnetic intensity (H) for an alternating magnetic field. As the magnetic field increases the ferromagnetic material becomes magnetized. As the magnetic field decreases, the decrease in magnetic intensity lags behind the magnetic field resulting in a hysteresis loop. This results in dissipation of energy in the form of heat. Usually, Joule losses are more significant than hysteresis losses in ferromagnetic materials.

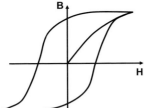

Figure 6.4 *Hysteresis loop for ferromagnetic materials*

The welding steps in implant induction welding are similar to those in other welding methods (see Chapter 2). Pre-welding preparation entails insertion of the gasket at the joint area and of the parts inside the coil. Induction heating of the gasket is done while the parts are pressed together against the gasket. In some cases, external pressure is difficult to apply and thermal expansion due to heating of the gasket and the joint is used to provide the needed welding pressure. Intermolecular diffusion occurs between the molten polymer in the gasket and the molten polymer on the part surfaces resulting in fusion bonding. For advanced composites and highly filled composites, squeeze flow and intermolecular diffusion are improved when the gasket is sandwiched between polymer films [2]. Finally, the parts are cooled and the weld is complete.

6.3   Process Description

The process of implant induction welding can be divided into the following steps:

- Gasket and part placement
- Application of pressure
- Induction heating

- Cooling under pressure
- Removal of parts.

Gasket materials for welding of thermoplastics are available in many forms (e.g., sheets, strands, tapes, and moldings) for easy placement in the joint area (see Fig. 6.5). In some applications it may also be possible to co-extrude the gasket with the part or use insert molding. Automated systems for dispensing of gasket material can also be used. As mentioned earlier, for advanced composites it is helpful to make the exterior of the gasket resin-rich or to sandwich it between two polymer films. Once the gasket is in place, the two parts are brought together and the assembly is placed in the machine.

Figure 6.5 Gasket preform types (Courtesy Ashland Specialty Chemical Company Emabond Systems)

Pressure can be applied in different ways depending on the application. For large parts, pressure may be applied by a press away from the coil and the joint area. In other cases, the coil may be embedded in polytetrafluoroethylene (PTFE) blocks or ceramic blocks and the blocks are used to apply pressure. For scanning systems, a pressure shoe might be used around the coil.

Once the parts are pressurized, the power is turned on and heating of the gasket is done for a predetermined period of time. When setting-up an application it is helpful to measure the temperature of the gasket. Standard thermocouples cannot be used in this case because of the magnetic field. Instead, optical temperature sensors or fluoro-optic sensors are utilized. During heating the parts are pressurized allowing the gasket to flow and fill the gap between the parts. At the same time, intermolecular diffusion occurs between the molten polymer in the gasket and the parts. After a predetermined heating time, the power is turned off and the parts are allowed to cool under pressure for a preset time. Then the parts are removed and the cycle repeats. It should be noted that some or all portions of the cycle can be automated.

6.4   Equipment

The components for an implant induction welding system include the generator, impedance matchbox, coil, press, fixtures, and gasket. The induction generator converts line electrical power to high-frequency power in the frequency range of 50 Hz to 8 MHz with power typically ranging from 2 kW to 5 kW and more. Figure 6.6 shows a 5 kW high-frequency generator.

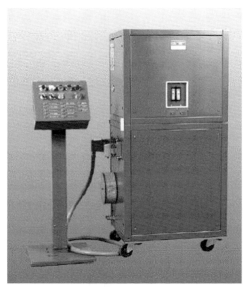

Figure 6.6 5 kW induction heating generator (Courtesy Ashland Specialty Chemical Company Emabond Systems)

The impedance matchbox can either be separate or incorporated into the generator. It is used to match the impedance of the loaded coil with the output impedance of the generator. This is sometimes referred to as tuning of the coil. Matching impedances is important for efficient and consistent operation of the system.

The work coil produces an alternating magnetic field. Since very high currents are generated in the coil it can get very hot. Therefore, hollow copper tubes are used to form the coil and cooling water is circulated through it. Coil design is very critical, as the coil geometry and electrical impedance can greatly affect the efficiency and rate of heating. Zinn and Semiatin [5] provide information on coil design and fabrication. Modeling of the coil and evaluation of the design is also possible by using many commercial finite element programs as well as custom coil design programs. During welding it is critical to maintain uniform proximity between the coil and the gasket otherwise non-uniform heating would result. This is especially important when operating at high frequencies. Figure 6.7 shows examples of multi-turn cylindrical, pancake, and spiral helical coils.

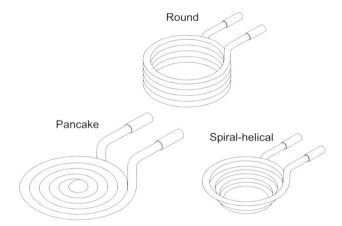

Figure 6.7 Examples of working coils

The fixtures provide the appropriate support for the parts to be welded and, when combined with a pneumatic press system, they apply the desired pressure (see Fig. 6.8). Fixtures for implant induction welding have to be carefully designed to avoid being heated by the electromagnetic field while at the same time being able to fully support and apply pressure to the parts. When necessary, PTFE or ceramic fixtures can be used to avoid heating of the fixtures by the magnetic field.

The gasket is a composite of the polymer and either an electrically conductive or a ferromagnetic filler and it is available in many forms (see Fig. 6.5). The choice of filler material is very important for the operating frequency range and for the weld quality. For lower operating frequencies it is usually better to select conductive fillers, while at higher frequencies ferromagnetic fillers are more effective. In some cases both conductive and ferromagnetic fillers are used. As mentioned earlier, the polymer in the gasket must be the same or at least a compatible material with the polymer used in the parts. Finally, it is important to remember that the gasket will remain imbedded at the interface. Therefore, filler materials that may corrode over time or in some other way may affect the performance of the weld should be avoided.

During welding, very high electric currents pass through the coils resulting in resistive heating of the coil. Therefore, it is necessary to have a cooling system circulating water through the coil so that it does not overheat and deform or melt. Figure 6.8 shows the water cooling system, which usually includes filters to ensure that clean water is used to avoid the clogging of small coils.

Figure 6.8 Implant induction welding system showing press, fixtures, and water cooling system (Courtesy Ashland Specialty Chemical Company Emabond Systems)

6.5   Joint Design

Joint design is also important for achieving the desired short- and long-term performance of the weld. Usually, the joint is designed in such a way that allows placing a preformed gasket at the weld area without it falling out during the process. This is usually accomplished by having a groove on the bottom part to hold the gasket. During welding, the gasket flows so as to fill the available space in the groove. This ensures complete contact between the gasket and the parts at the weld, thereby maximizing the weld strength. In most cases, the gasket and the weld perform better when loaded in shear. Figure 6.9 shows a few selected examples of joint designs. The flat-to-groove joint is quite effective when the plates are loaded in the plane or when the weld area is large. It does not perform well when loaded in tension. The tongue-and-groove joint is very strong because of the large contact area and because the gasket is loaded in shear. It does require that the parts be thicker near the weld. While not as strong as the tongue-and-groove, the shear joint requires less space and is sufficiently strong for many applications. Further sacrifice in strength for reduced space can be achieved by using the step joint. Both the shear and step joint are very effective for hermetic seals with good aesthetic finish.

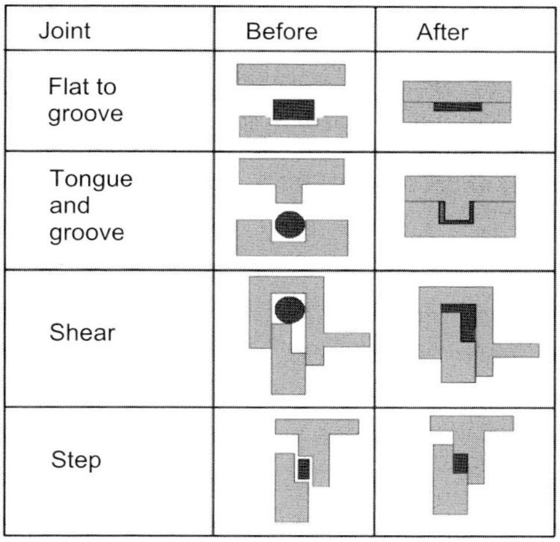

Joint	Before	After
Flat to groove		
Tongue and groove		
Shear		
Step		

Figure 6.9 Examples of joint designs for implant induction welding

6.6   **Material Weldability**

The weldability of thermoplastics by implant induction welding can be determined by considering the welding process parameters, the material compatibility, and appropriate selection of the equipment and especially the gasket. For implant induction welding of thermoplastics it is important to consider the following welding parameters:

- **Power** – Typically induction generators produce power in the range of 1 to 5 kW, although higher power generators are commercially available. Coil design and impedance matching are critical for efficient power transfer to the gasket.

- **Welding Time** – The power transmitted to the work coils determines the strength of the oscillating magnetic field produced. The more power generated the less time is needed to cause sufficient heating of the inductive polymer gasket. However, an appropriate balance between power and time must be achieved to allow sufficient heat flow to the parts without overheating the polymer in the gasket.

- **Welding Pressure** – Usually a pneumatic cylinder is used to apply a preset amount of pressure to ensure an even distribution of the gasket inside the joint.

- **Cooling Time and Pressure** – Holding under pressure will allow the parts to cool and resolidify. Hold times vary depending on the application, but are typically under 1s.

Evaluation of material compatibility and selection of appropriate gasket material has already been done for a wide range of thermoplastics. Table 6.1 shows a material compatibility chart for implant induction welding. Notice that in some cases, dissimilar material welding is possible. When the weldability of new materials needs to be determined, it is recommended to perform a systematic study using different susceptor and gasket materials and evaluating the effects of process parameters on heating rate and joint quality.

Table 6.1 *Material Compatibility for Implant Induction Welding (Courtesy Ashland Specialty Chemical Company Emabond Systems)*

	ABS	Acetals	Acrylics	Cellulosics	Ionomer (Surlyn)	Nylon 6.6, 11, 12	Polybutylene	Polycarbonate	Polyethylene	Polyphenylene oxide (Noryl)	Polypropylene	Polystyrene	Polysulfone	Polyvinyle chloride	Polyurethane	SAN	Thermoplastic polyester	TPE-copolyester	TPE-styrene bl. copolymer	TPE-olefine type
ABS	■		■					■								■				
Acetals		■																		
Acrylics	■		■					■								■				
Cellulosics				■																
Ionomer (Surlyn)					■															
Nylon 6.6, 11, 12						■														
Polybutylene							■				■									
Polycarbonate	■		■					■					■	■		■				
Polyethylene									■		■								■	
Polyphenylene oxide (Noryl)										■		■								
Polypropylene							■		■		■									■
Polystyrene		■						■	■			■				■				
Polysulfone								■					■							
Polyvinyl chloride														■						
Polyurethane															■					
SAN	■		■					■				■				■				
Thermoplastic polyester																	■			
TPE-copolyester																		■		■
TPE-styrene bl. copolymer									■										■	
TPE-olefine type											■							■		■

6.7   Applications

Implant induction welding is used in a wide range of applications including automotive, medical, food packaging, composite welding, and much more. It can be used on small and large parts and on a variety of materials.

Sealing of aseptic drink boxes is perhaps the largest volume application for implant induction welding (see Fig. 6.10). These boxes are made from complex multi-layered materials. One of the layers is aluminum foil that is used as an oxygen barrier to avoid discoloration and flavor change. The aluminum layer is inductively heated to melt the low density polyethylene on the interior of the package and to seal it. In this case, cycle time is very fast, typically under a half a second.

Figure 6.10 *Induction sealing of aseptic drink boxes*

Figure 6.11 shows a portable toilet made from polypropylene (PP) assembled by implant induction welding. Originally this toilet was manufactured using blow molding. By going to injection molding combined with implant induction welding it was possible to simplify the design, to include internal features, and to simplify the manufacturing process. The toilet has over 3 m of joint line length. In this case, a tongue-and-groove joint design was used to provide a hermetic seal with high consistency. With this joint design it is also possible to accommodate different styles using the same welding system.

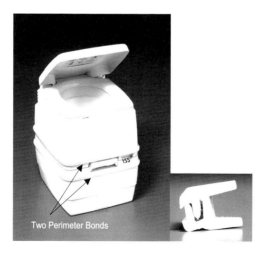

Figure 6.11 *Implant induction welding of portable toilet using tongue-and-groove joint (Courtesy Emabond Systems)*

Figure 6.12 shows station wagon seatbacks that are implant induction welded around the perimeter in addition to 12 individual spot welds for the individual seatback or 22 spot welds for the double seatback, respectively. These seatbacks are made from 40% by weight glass-reinforced PP. The perimeter bond is about 2.5 m long with 3-dimensional features. In this case, a step joint was used around the perimeter.

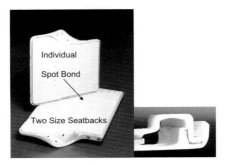

Figure 6.12 Implant induction welding of station wagon seatbacks using a step joint (Courtesy Ashland Specialty Chemical Company Emabond Systems)

Figure 6.13 shows a blood oxygenator that is made from polycarbonate and is implant induction welded. Eight sequential welds are made at the top, sealing each step (diameters ranging from 2.5 cm to 20 cm) individually. Implant induction welding enabled eliminating solvent bonding or adhesives to produce highly reliable pressure tight joints.

Figure 6.13 Implant induction welding of polycarbonate blood oxygenator (Courtesy Ashland Specialty Chemical Company Emabond Systems)

6.8   References

1. Chookazian, S.M., "Bonding with Electromagnetic Energy," *Plastics Assembly* (January 2000)

2. Benatar, A and Gutowski, T.G., "Methods for Fusion Bonding Thermoplastic Composites," *SAMPE Quaterly* (1986) **Vol. 18**, No. 1, pp. 35–42

3. Border, J. and Salas, R., "Induction Heated Joining of Thermoplastic Composites Without Metal Susceptors," Proceedings of the 34th International SAMPE Symposium (1989) pp. 2569–78

4. Schwartz, M.M., "Joining of Composite Materials," ASM International, Metals Park, Ohio (1994) pp. 35–88

5. Zinn, S. and Semiatin, S.L., *Elements of Induction Heating – Design, Control, and Applications* (1988) EPRI, Inc. and ASM International, Metals Park, Ohio

6. Davies, E.J., *Conduction and Induction Heating* (1990) Peter Peregrinus Ltd., London, United Kingdom

7. Sears, F.W., Zemansky, M.W., and Young, H.D., *University Physics – Part II* (1977) Addison-Wesley Publishing Company, Reading Massachusetts

7 Resistive Implant Welding

R. Wise, C. Brown, F. Chipperfield

7.1 Introduction

Resistive implant welding is a simple technique which can be applied to any thermoplastic polymer and almost any thermoplastic polymer composite. The technique involves the passing of a direct or low frequency alternating current through an electrically conductive implant placed between the parts to be joined. Electrical resistance (Ohmic) heating raises the temperature of the implant above the glass transition (T_g) or the melting temperature (T_m) of the thermoplastic polymer being joined, and a weld forms. Depending on the joint and application configuration, pressure either is generated internally through thermal expansion or is applied by the welding system. In applications where the pressure is internally generated, part tolerance (fit-up) is usually critical to achieving a good weld.

Resistive implant welding is often used where a high level of consistency is required and where applications can justify the cost of the implant (which in some cases can be significant).

Advantages and limitations of resistive implant welding include the following:

Advantages

- Simplicity – the technique is easy to implement
- The power input is easy to control via the electrical input (volts or current)
- The joints are assembled and then welded, so that molten polymer is never exposed – this reduces the risk of contamination
- Flexibility - almost any size of component can be welded (there will generally be a practical length limitation due to the safety implications of applying high voltages)
- Equipment required is generally inexpensive compared to other welding techniques such as vibration welding
- Welded parts can be disassembled for repair or recycling

Limitations

- Implants must be placed at the joint line
- Welding times can be relatively long compared to other techniques such as ultrasonic welding
- Welded parts contain the implant making them unattractive for recycling
- A reasonable level of skill is required to guarantee weld quality

Examples where resistive implant welding has been used successfully include welding of polyethylene gas pipes (Fig 7.1) and welding thermoplastic composites for aerospace applications.

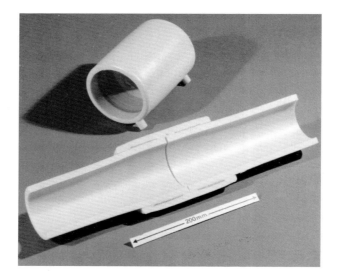

Figure 7.1 *An electrofusion coupler and a section through a pipe joined using an electrofusion coupler*

After a resistive implant weld has been made, the implant (which is usually metallic) remains at the joint line of the weld; this can be an important consideration if recycling of the components is likely in the future, because shredded components would contain a proportion of metallic material from the implants which would contaminate the recycled feedstock, limiting its range of application. The implant can also act as a stress concentration point due to the fact that it is a discontinuity in the weld zone. In addition, with metallic type implants, corrosion can sometimes be an issue depending on the application and service environment.

On the other hand, parts which have been joined using resistive implant welding can also be disassembled if sufficient electrical current can be re-applied to the implant and if the components can be separated. (Fig 7.2).

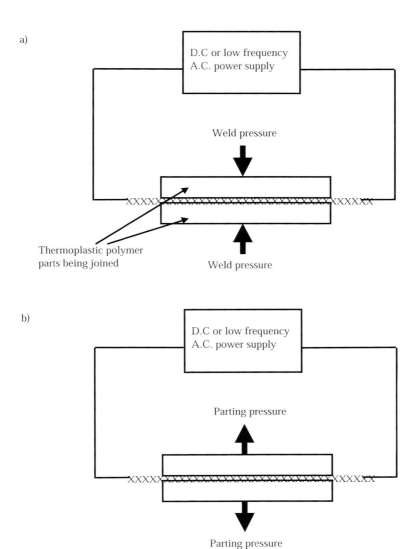

Figure 7.2 *Schematic showing a) resistive implant welding and b) disassembly*

7.2   Process Description

The use of D.C. or low frequency (<100 Hz) A.C. electricity to create a weld in thermoplastic polymer parts relies on the presence of an electrical conductor at the joint line. The conductor is necessary because most thermoplastic polymers are good insulators for electric currents in this frequency range.

A.C. is used where the cost of the power supplies is a particularly sensitive issue or where on-site power is provided by local diesel powered generators, e.g., for electrofusion welding of gas pipes in the field. D.C. is often used where accurate monitoring of the weld current is required or where current-controlled welds are made such as in the resistive implant welding of thermoplastic composite materials.

Resistive implant welding involves encasing an electrically conducting implant between the two parts to be joined and then resistively heating the insert by passing an electrical current through it. The temperature of the implant increases due to resistive losses and the thermoplastic surrounding it softens or melts as the glass transition temperature, T_g (in the case of amorphous thermoplastic polymers), or the melting point, T_m (in the case of semi-crystalline polymers), is exceeded. The choice of a suitable implant is very important and based on the following considerations:

- Cost
- Electrical resistance/unit length
- Geometry, e.g., joint width, and cross section of current carrier
- Length of flow for the polymer, i.e., how far the molten polymer must flow in order to contact the polymer material on the other side of the weld line
- Stiffness of the implant

Implants are often wires, braids or expanded sheets made from materials such as copper alloys, nickel alloys, or carbon fibers. The cross sections of the implants are important to the electrical resistance and to the mechanical properties of the welded joint. The main consideration when designing an implant is that adjacent wires are spaced sufficiently to allow the flow of molten polymer during welding. If this is not achieved, a polymer weld will only be achieved at the edges of the implant and a weak joint will result.

Because the implant forms part of the welded component, which will go into service, the effect of the implant on the mechanical properties of the component is very important. Figure 7.3 shows a cross section of a joint and highlights some of the main points of consideration.

The case shown in Fig. 7.3 is a generalized representation of a resistive implant weld and several variations are commercially practiced. Examples of these include implants molded into one component with a gap (see Section 7.5.1), and monofilaments of polymer included with the strands of the electrical conductor (see Section 7.5.2)

As shown in Table 7.1, the total welding cycle can be as short as one minute or may require several minutes.

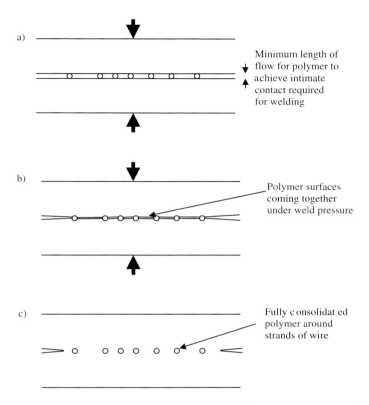

Figure 7.3 *Cross section through a weld in a thermoplastic polymer component as a function of time a) before welding, b) during welding – polymer flows around individual strands, and they are also pressed into the polymer by the weld pressure, c) after welding – intimate contact has been achieved, chain inter-diffusion has occurred and the material has frozen*

Table 7.1 *Typical Resistive Implant Welding Cycle*

Step	Typical time
Assemble parts	10 to 30 s
Attach electrodes	5 to 10 s
Current Applied	2 to 20 s
Hold time	2 to 20 s
Remove electrodes	5 to 10 s

In Fig. 7.3a, the gap between the two thermoplastic polymer surfaces before welding is defined. During welding, this gap is closed by a combination of three effects:

- Displacement of parts around the strands by the action of the weld pressure (if applicable)
- Flow of molten polymer around strands
- Thermal expansion of the polymer against the weld pressure (Fig 7.3b)

It should be noted that because most plastics have a high coefficient of thermal expansion, this internally generated pressure could be substantial. Figure 7.3c shows the section after welding, when the two components have come into intimate contact, polymer chain inter-diffusion has occurred and the material has frozen to produce a weld.

The shape and distribution of the individual strands is one feature of the welded joint which should be considered in any analysis of the mechanical properties of resistive implant welds, because it will affect the local mechanical stress distribution. It is unlikely that there will be significant adhesion between metallic strands and the polymeric material of the components being welded. Therefore, the volume of the strands can be considered to act as voids. For a known design stress, there will be an associated critical flaw size and it is important that the strand diameter does not exceed this.

Consistency of the welding technique is achieved by controlling the quality of the electrical and mechanical properties of the implants, and by the selection of a suitable power supply and delivery system, i.e., the leads and connectors conducting the current from the power supply to the weld. The process can also be monitored accurately by measuring the voltage dropped across the weld to help guarantee a consistent output. More sophisticated weld monitoring is also possible during weld process development or troubleshooting. In these cases, the temperature distribution generated in the weld can be monitored using either thermocouples buried in the weld or thermal imaging equipment.

7.3   Process Physics

The strength of implant resistance welds is a result of entanglement of polymer chains arising from their reptile like motion, as it is with all other polymer welding techniques [1, 2]. This mechanism is explained in more detail in Chapter 2. To achieve the right conditions to allow the polymer chains to entangle, correct welding temperature, time, and pressure have to be maintained.

Resistive implant welding basically involves supplying electrical energy to a resistor whose temperature increases sufficiently to melt a thermoplastic polymer (Fig. 7.4), and maintaining this temperature long enough to allow the weld to completely form.

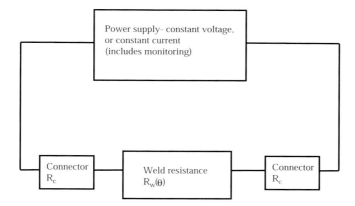

Figure 7.4 *Components of the basic electrical circuit (in block form) involved in resistive implant welding of thermoplastic polymers using direct current*

For simplicity reasons, Fig. 7.4 shows a welding configuration using D.C. although the low frequency A.C. form would be identical. For D.C., the power delivered to the weld resistance, R_w, would be:

$$\text{Power to weld} = I^2 R_w(\theta)$$

Where I is the current delivered and $R_w(\theta)$ is the resistance of the weld as a function of temperature (θ). The variation of the resistance of the implant with temperature can be used to monitor the weld and to ensure consistency from one weld to the next.

Another area of key importance is the provision of intimate contact between the two polymer parts during welding. This intimate contact can be achieved by the flow of molten polymer, by the thermal expansion of the polymer, and by the displacement of the implant into the softened polymer under the action of the weld pressure. The degree of flow depends on the melt viscosity, which is a function of parameters including temperature, flow rate, fillers, and the average molecular weight of the polymer. Thus, in order to achieve sufficient polymer flow, it is important to know the melt viscosity as a function of temperature so that the amount of pressure required to displace the implant can be estimated. Often, this is determined experimentally in a laboratory by varying the applied electrical power and pressure, and assessing flow by sectioning and mechanical testing of welded components.

The detailed analysis of a resistive implant weld providing distributions of temperature and pressure would typically be achieved by using a finite element model due to the complexity of the welding process.

Such a model has to accommodate factors including:

•   Complex joint geometry
•   Polymer thermo-mechanical properties

- Polymer thermal conductivity (a function of temperature)
- Polymer rheological properties (a function of temperature)
- Temperature dependent electrical properties of the implant

7.4   Power Supplies and Power Delivery

Power supplies for resistive implant welding are generally of the constant current or constant voltage type. Of these, the constant current type is preferable because it leads to the least variation in the power delivered to the implant, because most of the heating is achieved by resistive losses which, as discussed previously, are a function of current.

With reference to Fig. 7.4, a constant current power supply will always deliver the same current irrespective of variations in the resistance of connectors and leads. These variations will be automatically dealt with by a change in the voltage required to drive the demanded current. If the power supply delivers constant voltage, variations in the resistance of the implant will cause different amounts of current to be drawn. This will cause more or less power to be delivered to the load, leading to variations in weld properties. Despite these potential limitations in the performance of constant voltage power supplies, they are almost exclusively used for applications such as electrofusion welding of polyethylene pipes (Fig. 7.5). This is because they are less costly than constant current devices and for this application the small change in resistance of the metallic implant with temperature does not cause a large variation in the mechanical properties of the weld.

Figure 7.5 Polyethylene water pipes being welded using electrofusion

Modern resistive implant welding machines for applications such as welding polyethylene pipes would typically include the following features:

- A weld time selector
- A temperature compensation device (to adjust weld time to account for ambient conditions – important when welding outside)

Note: Weld pressure in this case is created by the thermal expansion of the polymer inside the implant fitting.

For applications where external pressure is required, such as welding thermoplastic composites, the welding machines would typically include the following features:

- Weld power ramp up time
- Weld time at peak power
- Weld power ramp down time
- Weld pressure control
- Time for which weld is held at pressure
- Weld cooling time under pressure

Note: The control over weld power during heating and cooling is required for the melting and recrystallisation of certain polymers (e.g., PEEK) to attain optimum mechanical properties.

7.5   Applications

There have been many applications of implant resistance welding because of its simplicity. Three examples will be discussed here: electrofusion welding of pipes, resistive implant tape welding, and resistive implant welding of advanced composites for aerospace applications.

7.5.1   Electrofusion Welding of Thermoplastic Pipes

7.5.1.1   Introduction

Thermoplastic pipes are widely used for the transportation of domestic gas, water, chemical effluent, and sewage. Low cost, flexibility, ease of installation, immunity to corrosion, and resistance to chemical attack are the most important advantages that polymer materials such as polyethylene (PE) and polypropylene (PP) offer over traditional pipe materials. The survival of PE and PP pipe systems following recent earth quake disasters in Japan and the western United States has seen these materials gain popularity in areas subject to ground movement.

When PE pipe systems were first used to distribute gas in the UK in 1969, the vast majority of joints were constructed using heated tool fusion techniques. The three original techniques of socket, saddle and butt fusion welding were successfully used for pipe joining, but at the cost of a high level of operator involvement requiring a considerable amount of ancillary tooling and equipment. Additionally, heated tool techniques were not suitable for joining different grades of material with dissimilar melt flow properties. While all heated tool methods were and still are operationally satisfactory, producing highly strong and reliable joints, a more modern joining technique was required. Electrofusion was developed as a system requiring minimal equipment, designed to improve efficiency and reduce the chance of human error.

7.5.1.2   The Electrofusion Welding Technique

This technique permits joining of pre-assembled pipes and fittings to be carried out quickly and efficiently with minimum equipment. . It also offers a number of practical advantages to the installer. It is easy to use for repairs, such as joining a replaced section of damaged pipe, and can be used where space is limited such as welding pipes in a trench. Electrofusion can also be used to join a wide range of resins, an important factor in its infancy, when there were many different grades in use for pipe applications.

7.5.1.3   Design of Electrofusion Fittings

The electrofusion welding process involves the use of a molded socket fitting containing an electrical resistive heating coil, see Fig. 7.6. This is basically an outer sleeve into which the two pipe ends to be connected fit. An internal stop prevents the pipe ends from meeting, thus creating a central cold zone between the fusion zones.

Electrofusion socket fittings can take the form of couplers, elbows, tees, and reducers. Additionally, there are saddle fittings designed to be fused to the surface of large diameter pipes to enable a small diameter service connection to be made, see Fig. 7.7

In Europe, the current range of PE electrofusion fittings covers pipe sizes from 20 mm to 630 mm.

An electrofusion socket fitting will typically include the following features:

- *A clearance* between the pipe outside diameter (OD) and the fitting inside diameter (I.D), to allow easier assembly and to accommodate tolerance variations.
- *Two heating coils* embedded close to the fitting bore. The heating coils are generally monofilar in construction with a continuous coil wound from one socket to another, thus enabling fusion of both pipe ends in one operation. In some cases, large diameter fittings have independent coils. The embedment of the coils, typically less than 0.25 mm from the fitting bore, protects them from dislodgement during assembly. This also places the heat source as close to the pipe surface as possible. Fusion indicators are designed into the fitting, which protrude from the outer surface when sufficient melt

pressure has developed at the joint line, giving a visual indication that molten material has formed.

• *An indication of the fusion and cooling times* to use. Manual fittings have this information embossed on the molding or displayed on an adhesive label, requiring the operator to enter the data into the electrofusion control unit (ECU) prior to welding. Automatic fittings have the data contained in a barcode or magnetic strip, scanned by light pen or card reader into the ECU.

a)

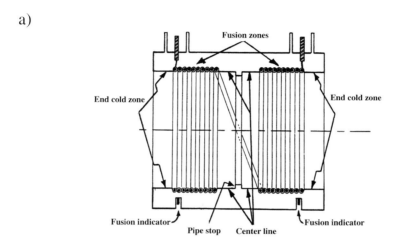

b)

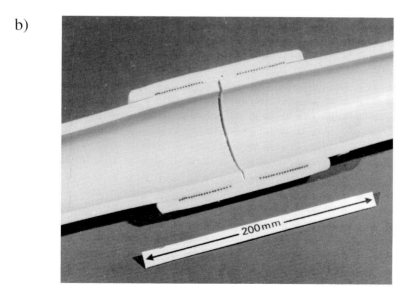

Figure 7.6 Cross section through an electrofusion fitting a) schematic b) through an actual gas pipe fitting

Figure 7.7 Electrofusion saddle fitting

7.5.1.4   Operating Voltage and Power Supply

A common voltage used by PE fittings across Europe is 39.5 V. This is considered a safe level for working in the field, keeping the current drawn to practical levels (maximum 50 A). Large diameter fittings of 355 mm and above are designed to run on 79 V, but operate at ±39.5 V relative to ground to minimize the risk of electrical shock.

Electrofusion control units (ECUs) are powered from an electrical mains or field generator supply having an output of 110 V and a rating of generally 3 to 3.5 kVA for 39 V fittings and 6 to 7.5 kVA for 79 V fittings. Most ECUs can operate in both automatic and manual mode to suit the type of fitting being fused. Many incorporate additional features such as data logger, joint counter, temperature compensation facility, and choice of operating language.

A recent development has been the introduction of a battery powered ECU, suitable for making a limited number of joints in service pipe sizes up to 63 mm. The requirement for an auxiliary power source is eliminated, giving total freedom of use when welding on site.

7.5.1.5   Preparation of Joints

Although semi-skilled operators with limited training may execute electrofusion welds, it has been demonstrated that in order to establish a consistent and structurally sound joint, it is necessary to follow a strict preparation procedure prior to making the weld. If the appropriate procedures are followed, contamination and disturbance effects that might inhibit the fusion mechanism will be minimized.

For electrofusion of pipes, Bowman [3] has demonstrated that there are three critical steps for joint preparation. First, the pipe ends must be squarely finished, and pushed against the central pipe stop in the fitting. This ensures that the central cold zones function to contain the melt and that none of the heating element wires are exposed (Fig. 7.8).

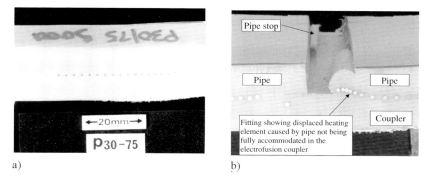

Figure 7.8 *Cross sections through the wall of welded electrofusion couplers showing a) a good weld and b) a weld defect caused by one pipe not being pushed fully into the coupler prior to welding*

Secondly, the pipe ends to be joined must be thoroughly cleaned and scraped to the correct depth, removing the oxidized outer surface. In electrofusion joining there is little or no relative movement of the pipe to the coupler during welding, hence any dirt or contamination on the pipe is retained at the joint, which prevents molecular diffusion and acts as a weld defect.

Finally, the pipes and fitting must be securely clamped to effectively eliminate relative movement during welding. This is required in order to contain the molten polymer at the fusion interface and to generate internal pressure, thus promoting better mixing of the polymer at the interface. In addition, this rigidity is important to ensure no movement takes place during cooling.

7.5.1.6   Stages in Electrofusion Welding

The joining process during electrofusion welding can be divided into three phases: (I) initial heating and fitting expansion, (II) heat soaking to create the joint and finally (III) joint cooling. Phases I and II are commonly termed "fusion time".

During the first phase, the polymer on the bore of the fitting expands on heating and fills the initial gap between the pipe and the coupler. This is possible because the volumetric expansion of polyethylene associated with a temperature rise from 20 °C to 250 °C is approximately 20%. Tests with varying "fusion times", together with the measurement of joint strength, carried out by Bowman [4], suggest that this first phase takes between 20% and 40% of the total "fusion time". Prior to the first phase, a pre-heat cycle is used with some designs of large diameter couplers. This cycle is designed to slightly expand the pipe OD and reduce any gap between the fitting bore and the pipe.

The energy supplied to the heating coil is distributed as heat to the pipe and fitting, thus creating a melt pool. The contained melt pool increases in volume on further heating to build up melt pressure. The combined action of the heat and pressure promote the formation of the fusion joint between the pipe and coupler. The weld strength increases appreciably between when the cycle time is between 40% and 80% of the total "fusion time".

The final phase of the joining process begins after the fusion time has elapsed and the joint is left to cool undisturbed for a specified time. The melt at the interface slowly cools and contracts. A slight distortion of the coupler, due to contraction stresses is sometimes observed. Electrofusion joints intended for pressurized pipe systems are given sufficient time to cool to ambient temperature prior to pressure testing.

7.5.2   Resistive Implant Welding Using Tape

The resistive implant welding of a wide range of components was facilitated greatly by the introduction of a braided tape containing tinned copper wire strands interwoven with monofilaments of thermoplastic polymer [5].

This approach was designed to reduce the flow path of molten polymer required to make a fully consolidated joint by minimizing the gap length as defined in Fig. 7.3. A schematic section through such a joint is shown in Fig. 7.9.

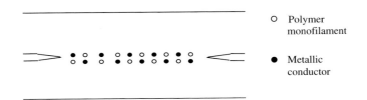

Figure 7.9 Schematic section through a resistive implant weld showing the distribution of metallic wire strands and polymer monofilaments

Because the polymer monofilament must match the polymer type of the components being welded, braided tape incorporating different polymer types is available.

Tapes having widths in the range from 4 mm to 15 mm are produced and the choice of width for any application will be made based on the geometry of the joint being made (Fig. 7.10a). For optimum results, the design of mating joint faces should be accommodate the welding tape, i.e., the joining method should be considered early in the design process. The parts to be joined should be placed in a jig or clamped with sufficient pressure during welding to ensure that full contact is maintained. The tape is sandwiched between the parts and electrical connections are made at each end. Low voltage (typically 30–50 V), high current electricity is applied across the tape; the exact parameters can be established by trials prior to production welding.

Generally, welding times are minimized for production and so the highest welding current possible without causing disintegration of the polymer is applied. For example, with 4 mm wide tapes, 150–180 A are applied. The exact value is a function of joint area, polymer type, thermal conductivity of the jig or clamps, and the welding pressure. Under ideal conditions, weld times are in the range 30 s to one minute.

a)   b)

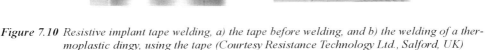

Figure 7.10 *Resistive implant tape welding, a) the tape before welding, and b) the welding of a thermoplastic dingy, using the tape (Courtesy Resistance Technology Ltd., Salford, UK)*

Two-part car bumpers were welded together using this technology with a feature designed to accommodate the braided tape molded into one of the bumper components prior to welding [6]. Resistive implant welding was originally selected due to the size of the components to be welded and because it offered a solution which was both quick and inexpensive. This application used 7 mm wide tape, with 80 A and 8 V, and took approximately 50 s.

Another application involved welding the hull of a dingy to the deck as shown in Fig. 7.10b. Other applications include welding vacuum cleaner casings, car batteries, loudspeaker housings, and high-load plastic pallets.

7.5.3   Resistive Implant Welding of Advanced Composites

In the early 1980s, a range of thermoplastic composite materials based on polyetheretherketone, polyetherimide, and other high temperature thermoplastic polymers began to appear. These materials were reinforced with carbon, glass, or aramid continuous fibers and were designed specifically for the aerospace industry.

One technique developed for joining them was resistive implant welding and implants based on carbon fibers and metal wires were developed. The carbon fiber based implants were lengths of unidirectional pre-impregnated composite tape, i.e., long carbon fibers oriented in one direction, surrounded by a matrix of thermoplastic polymer of the same type as the parts being welded (Fig. 7.11). Electrical contact was made to the carbon fibers either by burning off the polymer at the ends of the implant to expose the bare carbon fibers, or by penetration of the polymer with a metallic connector. The advantage of this type of implant is that good adhesion between the electrically conducting strands and the polymer is assured by the use of a composite as the implant. The mechanical properties of the weld were very good because the weld material (implant) was the same type as the parent material.

As insulators, the thermoplastic composites based on glass or aramid fibers were both easy to weld using implants of carbon fiber or metal wires. However, with the carbon fiber based components, a complication arose when some of the electric current applied to the implant deviated through the component rather than passing only through the implant, which is known as "shunting". The magnitude of this deviation was not consistent and led to some variation in the mechanical properties of welds made with this tequnique. In order to prevent shunting, it is important that the implant is electrically insulated from the parent material. This is typically achieved by placing an extra layer of the thermoplastic between the base material and the implant before the welding cycle is started [7, 8].

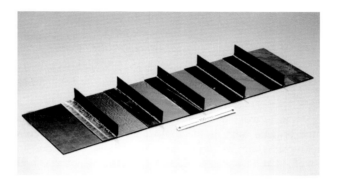

Figure 7.11 *A stiffened panel of carbon fiber PEEK (polyetheretherketone) composite manufactured by resistive implant welding the stiffeners to the panel using pre-impregnated tape as the implant*

An additional benefit of using carbon fibers as current carrying conductors in resistive implant welding was that the change in resistance was negative as temperature increased and it was relatively large. This facilitated weld monitoring by measuring volts dropped across the implant during the weld.

7.6   References

1. Wool, R. P., Polymer Interfaces: Structure and Strength, Carl Hanser Verlag, 1995

2. Wise, R. J., Thermal Welding of Polymers, Abington Publishing, 1999

3. Bowman, J., 'Fusion joining of cross-linking polyethylene pipe', Proceedings of 'Advances in joining plastics and composites', Conference held by TWI, Bradford, UK, 10–12th June 1991. Paper 13

4. Bowman, J., 'The assessment of the strength of electrofusion joints', Proceedings of '12th plastic fuel gas pipe symposium', Conference held by American Gas Association, Boston, USA, 24–26th, September 1991, p. 311–323

5. Hartley, P., Resistance Technology Ltd., Salford, UK, private communication, January 2002

6. Knight, R., Implant welding plastic bumpers, *Machinery and Production Engineering*, April 1983, p. 14–15

7. Eveno, E. and Gillespie, J.W. Jr., Resistance welding of graphite polyetheretherketone composites: An experimental investigation, *Journal of Thermoplastic Composite Materials*, **1**, October 1988, p. 322–338

8. Offringa, A., 'Manufacturing technology development for thermoplastic Fokker 50 main undercarriage doors', Proceedings of 14th International European Chapter Conference of SAMPE, Birmingham, October 1993

8 Ultrasonic Welding

David Grewell, Avraham Benatar, Joon Park

8.1 Introduction

Ultrasonic welding is a very popular technique for fusion bonding of thermoplastics and thermoplastic composites. Welding is accomplished by applying low amplitude (10 μm – 250 μm) high frequency (10–70 kHz) mechanical vibration to parts. This results in cyclical deformation of the parts, primarily at the faying surfaces (joining surfaces) and surface asperities. The cyclical energy is converted into heat – within the thermoplastic – through intermolecular friction. This is similar to the heating that occurs in a metal wire that is bent back and forth repeatedly, or in general, to the effect occurring when materials are subjected to cyclical loading. The heat, which is highest at the surfaces (because asperities are straining more than the bulk), is sufficient to melt the thermoplastic and to fusion-bond the parts. Usually, a man-made asperity in the form of triangular protrusion is molded into one of the parts to improve consistency (see Fig. 8.1). This protrusion, which is also called an energy director or concentrator, experiences the highest levels of cyclical strain producing the greatest level of heating. Therefore, the energy director melts and flows to join the parts.

Ultrasonic welding is frequently used for parts that cannot be molded as one piece due to complexity or cost. It is often used in mass production because the welding times are relatively short (only a few seconds). Ultrasonic welding is a flexible technique that can also be used in small lot size production - as long as the fixtures are designed to be flexible. It is applicable to both amorphous and semicrystalline thermoplastics. In some cases, the technique can even be used to join dissimilar materials [1].

Ultrasonic welding is one of the most common methods used in industry to join plastics. There is no one particular reason for its popularity but some of its advantages include:

- Speed (typical cycle times less than 1 s)
- Ease of automation
- Relatively low capital costs
- Amenable to a wide range of thermoplastics

Ultrasonic welding is usually divided into two major groups: near-field and far-field welding. Current industry practice, which is based on the most extensively used 20 kHz welding system, considers cases where the distance between the horn/part interface and the weld interface is less than 6 mm to be near-field welding (see Fig. 8.2). Far-field ultrasonic welding is used to describe cases where that distance is greater than 6 mm. At 20 kHz, the wavelength in the plastic component is between 6 and 13 cm depending on the specific polymer. Therefore, during near-field ultrasonic welding, the vibration amplitude at the weld interface is close to the amplitude at the horn face. For far-field welding, the amplitude of vibration at the weld interface depends on the ultrasonic wave propagation in the parts.

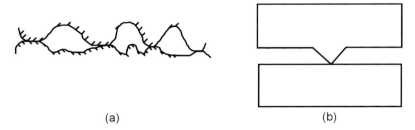

(a)   (b)

Figure 8.1 *Ultrasonic heating through cyclical deformation of (a) Surface asperities and (b) Man-made energy directors*

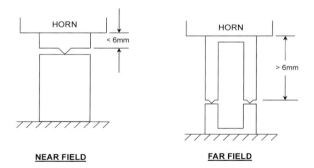

NEAR FIELD   FAR FIELD

Figure 8.2 *Industry accepted definition for near-field and far-field ultrasonic welding. (This defini-tion is applicable to operating frequencies near 20 kHz)*

Beyond simply welding plastic components, ultrasonic energy can be used to insert metal threaded bosses or other components into a thermoplastic part. When joining two dissim-ilar thermoplastics, which are not weldable, or when joining thermoplastic components to a second material such as metal, ultrasonic staking can be used (see Section 8.4 for details). Ultrasonic swaging is similar to ultrasonic staking where ultrasonic energy is used to melt and shape a polymer stud. Ultrasonic swaging is used in a variety of applications, such as to reshape the end of thermoplastic tubes or rods to a desired shape for connection with a coupler. Ultrasonic energy can also be used for many other applications including machining, cutting, sewing, degassing, separation of solids or of liquids, non-destructive inspection, imaging and more [2].

8.2   Process Description

Ultrasonic welding of thermoplastics can be divided into two modes of operation: plunge and continuous welding. Plunge welding is a discrete welding process: The ultrasonic tooling engages the parts to be welded/joined and retracts at the end of the process. In continuous welding, the ultrasonic tooling remains against the parts under a certain force or fixed gap and the parts (usually films or fabrics) are drawn through in a continuous fashion. The following section details these modes in more detail.

8.2.1   Plunge Ultrasonic Welding

Plunge ultrasonic joining of materials is common in welding, staking, insertion, and swaging. Plunge ultrasonic welding involves placing the parts to be joined into a fixture, initiating the cycle by applying the ultrasonic energy, and then removing the parts. Table 8.1 shows the general steps for plunge welding and the typical times for each step. Plunge welding processes can be either manual or automated. As shown in Table 8.1, total plunge welding cycle can be as short as a fraction of a second or as long as 20 to 30 s. It should be noted that cycle times as short as 0.1 s have been reported with special part loading systems.

Table 8.1 *Typical Ultrasonic Welding Cycle*

Step	Typical time
Load parts	Manually (3 to 5 s)/automated (0.1 to 3 s)
Press palm buttons/activate process	Manually (1 to 2 s)/automated (0 to 0.5 s)
Head lowers	0.25 to 2 s
Ultrasonic energy on	0.1 to 10 s
Hold time	0.1 to 10 s
Head raises	0.25 to 2 s
Remove parts	Manually (3 to 5 s)/automated (0.1 to 3 s)

Another convenient way to review the plunge welding process is by evaluating a typical cycle graph, such as the one depicted in Fig. 8.3. A typical cycle graph usually plots the data either at the start of the ultrasonic energy or just prior to the initiation of the ultrasonic energy. Thus, a typical cycle graph does not show when the cycle is initiated, usually by closing a pair of palm buttons or by a signal from a programmable logic controller (PLC). Once the actuator leaves what is known as its home position and travels until it makes contact with the part, which is referred to as the "down stroke." During this travel, it is possible to initiate the ultrasonic energy at any given distance of travel, which is called pre-triggering. Pre-triggering is usually necessary when high gain and/or a large horn is used, which requires a significant amount of power to run or start the horn to vibrate. Figure 8.3 shows a cycle where the ultrasonic vibration is initiated once a particular force – known as trigger force – is achieved (in this case approx. 30 N). As seen in the graph, the force rises as a function of time until it reaches a steady-state value, which corresponds to the pre-set welding force. This is a typical force build-up curve for pneumatic driven actuators. Once the ultrasonic energy is initiated, it is seen that the power quickly rises, because enough power must be delivered to system not only start the welding processes, but also to initiate the vibration of the stack (transducer/converter, booster, and horn). The magnitude of the power spike depends on the stack and application and is often the highest power dissipation during the entire cycle. The distance curve indicates that at this point the samples begin to melt and collapse (displace). After some length of time, which is application-dependent, the distance curve reaches a steady state condition in which the curve becomes relatively linear with respect to time. When this stage is reached, the entire faying surfaces are usually molten and the ultrasonic energy is discontinued. It should be noted that it is common industry practice to discontinue the ultrasonic energy well before the displacement reaches a linear slope, because the joint strength meets or exceeds the application requirements, although maximum strength may not be reached by early termination. After discontinuation of ultrasonic energy the tooling usually remains in contact, under a pre-set force known as the hold force, for a time usually set to half the weld time (ultrasonic-on time). This phase of the cycle is known as the hold time. Early during the hold time a small amount of collapse may occur due to squeezing of the residual melt prior to solidification of the interface. After the hold cycle, the ultrasonic tooling retracts and returns to the home position, a motion known as the "up-stroke". In some applications, during or after the up-stroke, the ultrasonic energy is activated for a short period of time (less the 0.5 s). This is known as "post weld burst," and it is often used to free the tooling of any material that may have transferred from the parts to the tooling during the welding cycle.

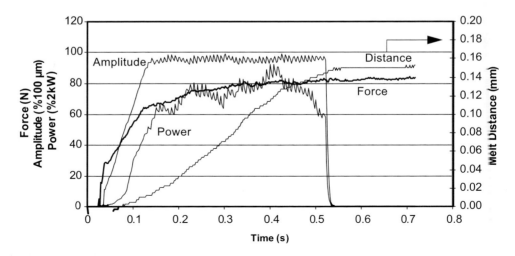

***Figure 8.3** Typical cycle graph for plunge ultrasonic welding*

8.2.2   Continuous Ultrasonic Welding

Continuous ultrasonic welding is almost exclusively used with film and fabric applications. In this case, the materials to be joined are placed between the horn and the fixture and fed through at a particular rate. This type of application is similar to sewing and a lot of continuous ultrasonic welding equipment is designed to look and function like sewing machines. However, in large continuous welding applications, the equipment is usually much larger and the design can be quite different from a sewing machine.

During continuous ultrasonic welding the material experiences similar forces and vibrations to those experienced during plunge welding. Usually the horn or fixtures have rounded edges at the incoming location for the material. Therefore, as the material enters the space between the horn and fixture, the force builds up gradually and the ultrasonic vibration that is experienced by the material also increases gradually until the steady state force is reached. As the material exists the horn/fixture gap, a hold pressure is applied during cooling by adjacent welded material, which is already cooled and solidified. In some cases, rollers may be used to apply the hold pressure.

8.2.3   Process Control

The type of control that is used during the ultrasonic welding process depends in part on the type of welder that is used. A wide range of equipment is available; starting from simple hand-held welders to fully microprocessor-controlled welders with the ability to interface with personal computers, which may even have built-in statistical process control (SPC) software. When selecting an ultrasonic welding machine for a particular application, the level of process control that is necessary becomes a primary consideration. For example, to create spot welds between two sheets of polyethylene, it is possible to use a simple hand-held welding system that has no automated controls. On the other hand, for a medical application where control of joint quality and final dimensions are critical, it is necessary to use a microprocessor-controlled welder with SPC software.

There are many control modes of operation as shown in Table 8.2. The simplest and most common mode is the "time mode." In this mode, the ultrasonic energy is activated for a pre-set length of time regardless of the other parameters. Another popular mode is the "energy mode." This mode usually requires a microprocessor-based controller that can measure and record the power delivery during the weld cycle in order to calculate the energy (the integral of power over time). In this mode, the ultrasonic vibration remains on until a pre-programmed amount of energy is delivered into the parts regardless of other parameters. Similarly, in the collapse mode, the ultrasonic vibration remains on until a final collapse or travel distance is reached. In the peak power mode, the ultrasonic vibration is kept on until the instantaneous power exceeds a preset peak power level. Ground detect mode is used to control the final dimensions of the welded part by continuing the ultrasonic vibration until the horn and fixture make contact. In this mode, it is common for the fixture to be electrically insulated from the base of the machine. A variation of the ground detect method is used in continuous ultrasonic welding to control the gap between the horn and fixture. In this case, a servo-actuator is used to move the horn relative to the fixture to maintain a constant resistance measurement.

Table 8.2 *Typical Modes of Operation for Process Controls*

Mode	Details of mode	Level of control
Time	Weld for preset length of time	Basic system
Energy	Weld until preset energy is delivered into part	Microprocessor based controller
Collapse	Weld until the parts have collapsed to a preset value	Microprocessor based controller and encoder
Peak power	Weld until a preset power level is achieved	Microprocessor based controller
Ground detect	Weld until the horn make electrical with the fixture	Microprocessor based controller

It should be noted, that with any of the above-mentioned process control modes, it is possible to use one control parameter and define limits on the other parameters for quality assurance reasons. For example, when welding in the time mode, it is possible to set the minimum weld energy and/or a maximum power level. In this case, if either of the secondary parameters (energy and/or peak power) fails to meet the pre-set conditions at the end of the weld time, the controller can either alarm the operator or a PLC to enable inspection or rejection of the part.

8.3   Physics of Process

8.3.1   Near-Field Ultrasonic Welding

Near-field ultrasonic welding is a complex process that consists of six distinct yet highly coupled sub-processes [3]:

1. Mechanics and vibrations of the parts,
2. Viscoelastic heating of the thermoplastic,
3. Heat transfer,
4. Flow and wetting,
5. Intermolecular diffusion, and
6. Cooling and re-solidification.

The mechanics and vibrations model of the parts, the fixture, and the welder determine the strain distribution within the components and at the faying surface. From the strain distribution, it is possible to determine the heating that results in the parts. In order to form a good weld – rapidly – it is necessary to concentrate the ultrasonic energy at the weld zone. In the case of energy directors, this is accomplished by forming V shaped protrusions or energy directors on the part surface (see Fig. 8.1b). In some cases the joint design has an interference fit between the parts, which is called a shear joint, (see Fig. 8.4).

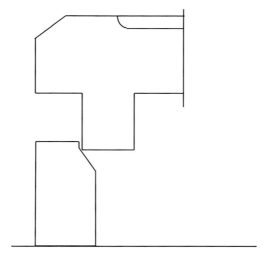

Figure 8.4 Typical shear joint for ultrasonic welding

Viscoelastic heating of a polymer can be understood by examining the mechanical behavior of these materials. Polymers exhibit a viscoelastic behavior, i.e., they behave as a combination of an elastic (spring like) material and a viscous (dashpot like) material. As shown in Fig. 8.5, for a spring subjected to cyclical loading the force and displacement are in phase and the force and velocity are 90° out of phase.

Therefore, a spring does not dissipate energy; the energy that is supplied to elongate the spring during the first half of the cycle is recovered during the second half of the cycle. Figure 8.6 shows that for a dashpot that is subjected to cyclical loading the force and displacement are 90° out of phase and the force and velocity are in phase. Therefore, a dashpot dissipates all of the energy that is provided to it.

For a viscoelastic material subjected to cyclical loading, the stress (force) is out of phase with the strain (displacement) by some angle that ranges between 0° and 90° (see Fig. 8.7). Therefore, for the viscoelastic material, some of the energy input is recovered (through the spring-like behavior) while some of the energy is dissipated (through the dashpot-like behavior). Therefore, for the viscoelastic material the dissipated energy will result in heating and melting of the polymer during ultrasonic welding.

The mechanical behavior of a viscoelastic polymer is represented in the form of a complex dynamic modulus, $E^* = E' + iE''$. Where E' is the storage modulus and represents the ability of the material to store energy like an elastic material. E'' is the loss modulus and represent the ability of the material to convert mechanical energy into heat like a damper. One might also use the loss tangent (tan δ) to represent this behavior as a ratio,

$$\tan \delta = \frac{E''}{E'} \qquad\qquad (8.1)$$

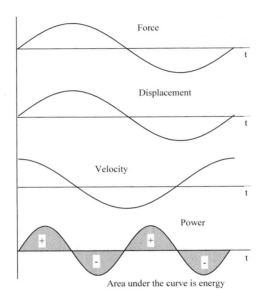

Figure 8.5 Cyclical loading of spring

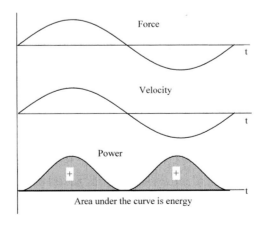

Figure 8.6 Cyclical loading of a damper or dashpot

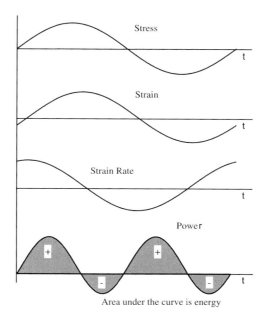

Area under the curve is energy

Figure 8.7 *Cyclical loading of a typical viscoelastic material*

For most metals, the loss tangent is around 0.001 or less while for polymers it is in the range of 0.001–0.5. Therefore, polymers can dissipate a significant amounts of energy during ultrasonic vibration. The average internal heat generation ($\dot{Q}_{avg}$, energy dissipated per unit time and per unit volume) is

$$\dot{Q}_{avg} = \frac{\omega E'' \varepsilon_0^2}{2}$$ (8.2)

Where ω is the operating frequency (rad/s) and ε_0 is the strain amplitude, which in most cases is directly proportional to the vibration amplitude. Unfortunately, the dynamic loss modulus of the polymer is both frequency and temperature dependent and it is very difficult to measure at the excitation frequency for ultrasonic assembly. A variety of techniques may be used for the measurement of the dynamic properties of polymers [4, 5]. Low frequency measurement is usually favored because of commercially available equipment, and because inertial effects are negligible. In addition, the commercial equipment incorporates environmentally controlled chambers and computer interfacing, which allow for time-temperature superposition to estimate the dynamic properties at higher frequencies.

As the energy directors get hotter from the dissipation of vibrational energy, heat is conducted from the energy directors into the (relatively) cooler parts. Heat conduction is much greater than the convective heat loss to the air [3]. Once the temperature in any portion of an energy director exceeds the melting temperature, the energy director will start to flow – primarily as a result of the forces applied by the welder. This complex flow process is described as a squeezing flow under the influence of both static and dynamic loading [3]. The flow of the molten energy director combined with additional ultrasonic heating results in melting of the part surfaces and thus achieves intimate contact at the interface. The diffusion of long polymer chains across the bond interface – and entanglement of these chains – is what gives the ultrasonic bond its strength. The reptilian diffusion theory [6, 7] is used to describe intermolecular diffusion and chain entanglement. After the ultrasonic vibration is stopped, the parts begin to cool while additional squeeze flow continues under the static pressure until finally the molten polymer at the interface solidifies. It is during this stage that the final microstructure, residual stress, and distortion are determined.

8.3.2   Far-Field Welding

Far-field ultrasonic welding is used to describe cases where that distance between the horn/part interface is greater than 6 mm. At 20 kHz, the wavelength in the plastic component is between 6 and 13 cm depending on the specific polymer. Therefore, during near-field ultrasonic welding, the distance from the horn/part interface to the weld interface is small compared to the wavelength, and the vibration amplitude at the weld interface is nearly equal to the amplitude at the horn face. For other operating frequencies, the industry accepted definition of a horn-to-interface distance of more than 6 mm, is not suitable. For example, at higher frequencies, the wavelength is shorter and a distance of 6 mm becomes too large to neglect wave propagation effects. Menges and Potente [8] studied the effect of part length on ultrasonic welding. They showed that the length of both the bottom part and top part relative to the wavelength are important depending on the type of joints used at the interface. Benatar and Cheng [9] propose modeling the wave propagation in the viscoelastic polymer to estimate the average or effective amplitude of vibration at the joint interface. The heat generation and squeeze flow can then be estimated using a Voigt-Kelvin model for the energy director with this effective amplitude of vibration.

8.4   Equipment Description

Ultrasonic plastic welding equipment is designed and built to operate at a particular frequency ranging from 10 kHz to 75 kHz. Each frequency has benefits and limitations. At lower operating frequencies, it is possible to deliver higher power. However, higher operating frequencies tend to promote less damage and also produce less audible noise. For theses reasons, most commercial equipment operates between 20 and 40 kHz. Table 8.3 shows a general trend of the power capabilities of most commercial equipment for various frequencies and the major benefits and/or limitations with the particular frequency. Since all commercial machines are designed to resonate at a particular frequency, it is not possible to modify the operating frequency of a machine, particularly the operating frequency of the stack (converter, booster, and horn) and power supply. However, actuators can be modified to operate at different frequencies. It should also be noted that Table 8.3 is based on the concept of using a single converter. Current technologies using multiple converters allows for relatively high power capabilities per horn (+2,000 W/converter). It should also be noted that machines operating above the audible human frequency (~18 kHz) still produce noise due to sub-harmonic modes of vibrations developed within the parts. This effect is minimized at higher frequencies. General guidelines for equipment selection are presented later in this section.

Table 8.3 *Comparison of Operating Frequencies for Ultrasonic Welding of Plastics*

Operating Frequency	Typical power capabilities/converter	Typical maximum converter output	Comments
10 to 20 kHz	6000 W to 3000 W (continuous)	40 to 20 (μm_{pp})	Relatively noisy, high power
20 to 30 kHz	3000 W to 1000 W (continuous)	20 to 10 (μm_{pp})	Moderate noise and power
40 to 75 kHz	1000 W to 400 W (continuous)	10 to 5 (μm_{pp})	Very quiet, low power

Most ultrasonic welding equipment also falls into one of two design concepts;

(1) Modular systems and
(2) Integrated systems.

In modular systems, the controls and power supply are separated from the actuator. These systems are commonly used when high levels of controls are needed as well as by OEM's that need packaging freedom. The integrated systems tend to be less expensive but have fewer controls. Figure 8.8 shows photographs of the two systems.

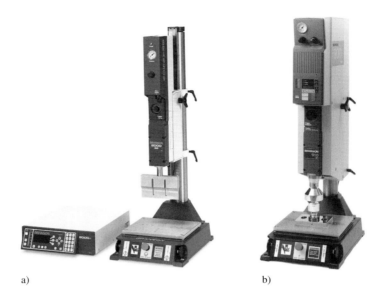

Figure 8.8 *Typical ultrasonic welding machine (plunge type), (a) modular system and (b) integrated system (Courtesy Branson Ultrasonics Corp.)*

There are seven basic components of an ultrasonic welding system. Below is a brief description of each of these components.

Power Supply

The power supply converts line voltage to high frequency power. Figure 8.9 shows a typical equivalent circuit for a commercial power supply. It is seen that line current is rectified from AC to DC. The DC current is then switched on and off at the operating frequency of the system. This switch is usually controlled by a system that assesses the power supply and stack to insure that they are properly matched to the resonant frequency to assure a high operating efficiency. In older systems, the power supply would need to be tuned to the mechanical resonant frequency of the stack much like tuning a radio to a particular station. In newer systems, the power supply will automatically tune itself as well as track frequency changes due to such influences as operating temperature of the stack. The power supply also contains self-protection circuits and power meters. Newer models have amplitude regulation to assure that outside influences such as line voltage fluctuations and power draw do not affect the converter output amplitude. The power supply is usually contained within the same housing as the controller.

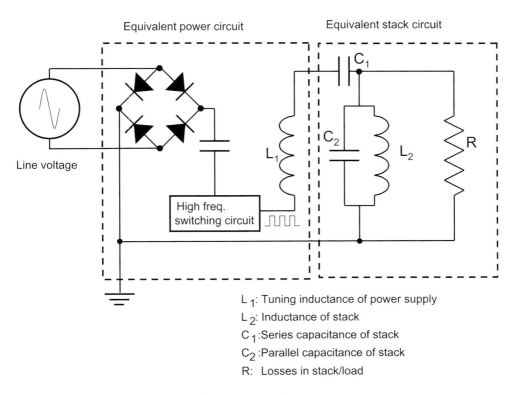

Figure 8.9 *Typical equivalent circuit for a power supply*

Controller

This component controls and monitors the system and interfaces with the user as well as PLCs if applicable. As discussed in Section 8.2.3, some of the aspects monitored by the controller include: weld distance (collapse), power, peak power, energy, force, and time. There is a wide range of controllers available on the market from simple time-based systems to personal computer-based systems with touch screen technologies. The selection of the type of controller is usually based on the application. For example, it is relatively standard in the medical industry to select a system that can monitor and record as many parameters as possible due to product liability. Or, if the application is a shear joint design, staking, or insertion, it is typical to select a controller that can weld to a particular distance (weld by distance), which requires the controller to be interfaced to an encoder. When welding film and fabric applications it is often useful to have a controller that can weld to a particular power level. Newer controllers even allow the amplitude to be varied during the weld cycle.

Converter

The converter is a motor, which is used to convert electrical energy to mechanical energy (vibrations). It is usually rated in terms of operating frequency and power. A typical converter can output peak-to-peak amplitude of as much as 20 to 25 μm_{pp} at 20 kHz. This is usually accomplished by either piezoelectric or magnetostrictive components. Most commercial ultrasonic welders use piezoelectric transducers because they are more efficient when operating at resonance. Figure 8.10 shows a photograph of a 20 kHz piezoelectric converter that has been partially opened. In the case of the piezoelectric system, relatively high voltages usually drive the converter while relatively high currents typically drives the magnetostrictive converters. It should be noted that the power restrictions of a converter are dictated by its frequency. Since the converter is usually designed to resonate at a particular frequency to assure high efficiencies (+90%), and the converter is also usually designed to be a half a wavelength in length, the converter length is fixed. This limits the physical size of the converter, which in turn limits the surface area and possible heat flow out of the converter. Thus, any internal losses of the converter can cause heat build-up, which can cause tuning and stress problems. Thus, the length of the converter decreases with increasing frequency reducing the available surface area for heat dissipation and thereby limiting the possible continuous power level of the converter. Table 8.3 shows this trend of decreasing power capability of the welding equipment with increasing operating frequency.

Figure 8.10 Interior of 20 kHz converter (Courtesy Branson Ultrasonics Corp.)

Booster

The booster is used to amplify or reduce the vibration amplitude. The converter is connected to one end of the booster and the horn is connected to the other end of the booster. Another important function of the booster is to provide a clamping location at its nodal plane/point (usually located near its center). A metal ring located around the body of the booster at the nodal point (plane) acts as a clamping ring and is usually painted in a particular color to indicate its gain. Most manufacturers provide a table showing the correspondence between color and gain as well as a clear labeling of the gain on the booster. Figure 8.11 shows a photograph of boosters with various gains. Manufactures usually offer boosters with different stiffness levels for the clamping ring. In selected applications, the stiffness of the booster can have significant influence on weld quality and weld data reported by the controller. In standard units, the booster clamping ring is at the nodal point and in order to minimize transmission of vibrations to the clamping structure, it includes rubber O-rings. Using O-rings with a square cross-section can increase the stiffness of the clamping ring. Further increase in stiffness of the clamping ring is possible by using direct metal-to-metal contact or one-piece solid boosters that include the clamping ring. These designs tend to be more costly and can introduce problems with long bar horns.

Figure 8.11 Boosters of various gains (Courtesy Branson Ultrasonics Corp.)

Horn

The horn has two main functions: (1) to further increase the amplitude of vibration (similar to the booster) and (2) to apply the ultrasonic energy to the work piece. It is usually ½ wavelength long and is usually machined from aluminum or titanium. Figure 8.12 shows the geometry, stress distribution, and amplitude distribution in a few standard horn designs: step, exponential, and catenoidal. It is seen that by changing the shape of the horn, it is possible to change the stress distribution within the horn as well as its gain. In the example in Fig. 8.12, the step horn is easier to manufacture but it has a relatively low gain and it has high internal stresses. On the other hand, the exponential horn has a high gain and relatively low internal stresses, a more desirable combination.

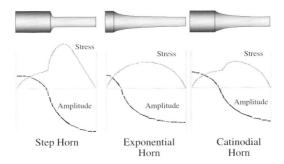

Figure 8.12 Geometry, amplitude distribution, and stress distribution in a step, exponential and catenoidal horns

In high wear applications, such as cutting or insertion, the horn can be machined from steel or have a carbide insert brazed to the face. In these applications, there is usually a conservative maximum amplitude restriction to limit the stresses induced on the horn. Often the horn can have complex designs, such as nodal plungers. A nodal plunger often consists of a piece of rubber or springs that are mounted to the horn at the nodal point and push against one of the parts being welded. The plungers are often added to a horn to reduce off-modes of vibrations that can develop in the parts during the welding cycle. Other horn designs may incorporate a part alignment device that is again usually mounted at the nodal point of the horn. Horns may also have knives mounted on their faces for cutting applications. For very wide horns, slots are machined into the horn to produce a more uniform vibration amplitude at the horn face. Figure 8.13 shows an example of a 10 cm wide blade horn with and without slots. By optimizing the size and location of the slots, the amplitude of vibration on the face of the horn can be uniform to within less than 5 %.

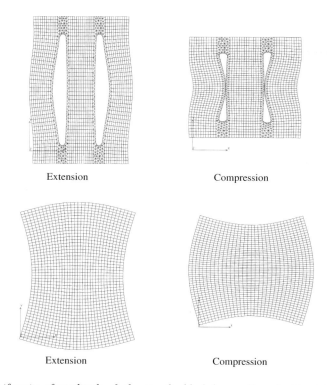

Extension   Compression

Extension   Compression

Figure 8.13 Uniformity of amplitude of vibration for blade horn with and without slots

Actuator

The actuator holds the stack assembly (converter, booster, and horn) and brings the stack into contact with the part. Most actuators are based on pneumatics, however, there are some on the market that are servo driven. Like controllers, there is a range of actuators available on the market offering different control levels. The simplest levels have mechanical trigger mechanisms (point of initiating the weld cycle) while others may have calibrated load cells to accurately measure the force level. Others may also have encoders that allow the travel distance to be reported during the weld cycle, or the travel distance may be used as a control parameter as discussed in Section 8.2.3. As with the controller, the selection of the actuator is usually based on the type of application as well as its specific requirements.

Fixture

The fixture is often considered the "cheap" component and therefore it does not usually get the attention it deserves. The fixture serves two functions:

(1) Holds and secures the parts and
(2) Provides location of the parts.

These two functions are often very critical in applications and thus it is important that care is taken to have properly designed and manufactured fixtures. Fixtures are usually machined from aluminum or steel. They can also be poured from urethane-like materials when a complex geometry is required. In addition, the fixture can have moving sub-components for easy loading of the parts and unloading of the welded assembly. Moving sub-components are sometime needed when welding parts with a shear joint. With shear joint designs, in order to prevent part deflection because of the interference fit of the parts, the fixture must hold the part so snugly that without an un-clamping mechanism, the part could not be removed from the fixture at the end of the cycle. In many fixture designs, heat dissipation, or the removal of heat from the parts can be an important aspect. In these cases, it is common to design the fixture from a material with high thermal conductivity, such as aluminum. In other cases, part marking can be issue, and the fixture is designed from a material with a low coefficient of friction, such as Teflon.

8.4.1   Selection of the Proper Equipment

Equipment selection is a complicated process where many factors have to be considered simultaneously. Therefore, this section is aimed at providing some general information to aid in the selection of equipment which should be done in close cooperation with ultrasonic welding experts and/or equipment suppliers. Table 8.4 shows some of the primary factors along with simplified guidelines to aid in equipment selection.

When considering equipment selection, one of the main considerations is operating frequency. Most equipment on the market covers the range of 20 to 40 kHz with some equipment covering 10 to 70 kHz. It is usually not cost effective to modify the operating frequency of a machine because the stack (converter, booster, and horn) must be tuned to resonate at a particular frequency and the power supply is usually designed to operate in a specific narrow frequency range. Usually the only component that can be modified to operate at a different frequency is the actuator and this is usually accomplished with a coupling ring that permits mounting of smaller converters and boosters. The main factor that determines the proper operating frequency for an application is the part size. As part size increases, the operating frequency of the equipment should be decreased, because usually larger parts require more power, which can only be achieved at lower frequencies. In addition, larger applications usually require larger horns, which can only be properly designed to operate at lower frequencies. For smaller parts or for parts that have very fine features and/or thin walls it is beneficial to use lower amplitudes of vibration to minimize resonant vibrations of the parts, which in many cases can lead to damage of the parts. As shown in Eq. 8.1, it is possible to increase the power dissipated in the polymer by increasing the operating frequency and/or increasing the vibration (strain) amplitude. Therefore, in many cases, the only possible way to obtain a combination of lower vibration amplitude together with substantial power levels means using higher operating frequencies.

Once the operating frequency is determined, the power capacity is usually considered. For each operating frequency, there are usually several available levels of power capacity. For example, at 20 KHz, most equipment suppliers offer ultrasonic welders with power ranging from 1 to 5 kW. In most cases, testing in an application lab of an equipment supplier, a research organization, and/or a university can be used to establish the power requirements. If it is not possible to pre-determine the power requirement theoretically, therefore it is best to select the highest power available to assure that the equipment can meet the requirements of the application.

Table 8.4 *General Guidelines for Equipment Selection*

Part size: Characteristic part size	
Less than 3 in.	All frequencies (500 to 1000 W)
3 to 15 in.	20 kHz and below (1000 to 3000 W)
Larger then 15 in.	15 kHz and below (over 3000 W)
Application: rigid vs. flexible	
Rigid	Plunge equipment
Flexible with closed pattern	Plunge and continuous system/actuator
Flexible with continuous pattern	Continuous
Welding distance: near vs. far-field	
Near field	All frequencies
Far field (amorphous)	All frequencies
Far field (crystalline)	Depends on part size, usually 20 kHz and below

Also, when selecting equipment it must be determined if the application is best suited for plunge or continuous mode of welding. If the application is a sewing type application, many equipment manufacturers offer standard machines of this type. They also offer many different standard rotary anvils that have different patterns. Many of these anvils can cut and seal as well as weld. If the application requires a very wide continuous seam (ranging from 5.2 cm to more than 122 cm), specialized equipment must be designed.

Other equipment considerations include control levels, amplitude regulation, amplitude profiling™ [10], and SPC capabilities. For some applications, these options can increase the weld quality and consistency. Again, testing in an application lab can help determine which control options are necessary. It should also be noted that with most equipment manufacturers it is possible to add some options after the original purchase of the welding equipment.

8.5   Joint and Part Design

8.5.1   The Importance of Part Design

Proper part and joint design, which will be defined shortly, are often the most important features to successful ultrasonic welding. With proper part and joint design even difficult applications, such as a part molded from PE requiring a hermetic seal, can be accomplished consistently. However, with poor designs even simple applications can be nearly impossible to weld. Because of the importance of a properly designed part, it is highly recommended that any proposed design be reviewed by a person who has a significant amount of experience in the field of ultrasonic welding of plastics. Usually equipment suppliers have a wide range of experienced applications engineers that can review drawings of proposed parts and recommend changes that might increase the likelihood of success. Consulting an expert after a mold is cut and problems have already arisen, can be costly. The best rule of thumb for part design is to do it right the first time. Because of the possible complexity of a part, the following sections are not intended to replace the review of a proposed design by an expert, but are intended to give the reader a foundation on part and joint design.

8.5.2   Design Factors

For any application, there are two major factors in design:

* Joint design, and
* Part design.

Focusing on joint design first, the two major categories to be considered are:

* Energy director
* Shear joint.

8.5.3   The Energy Director Design

The energy director is a small triangular protrusion molded into one of the parts along the weld joint line, where ultrasonic energy is concentrated for heat generation. Depending on part geometry and resin type, there are a variety of energy director designs. Joint designs with energy directors work best with amorphous resins but they are also very effective and frequently used with semi-crystalline polymers. Figure 8.14 shows the basic energy director design and general guidelines for dimensions. Table 8.5 lists the advantages and disadvantages of the energy director design.

Table 8.5 Advantages and Disadvantages of Energy Directors

Advantages	Disadvantages
Low energy dissipation	Limited weld strength with some materials
Minimal part marking	Some weld flash
Shorter cycle times	

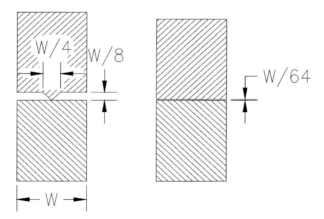

Figure 8.14 Basic energy director joint design (before and after welding)

Many of the disadvantages can be eliminated or minimized by using variations of the energy director design, such as using flash traps to keep the flash from flowing into visible areas. A typical flash trap with an energy director is shown in Fig. 8.15 along with many other popular variations.

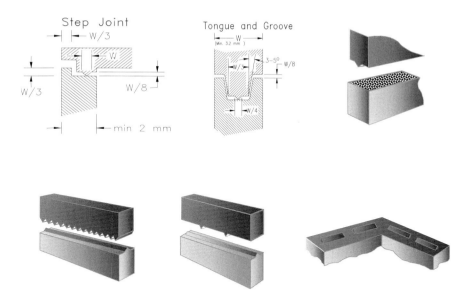

Figure 8.15 *Other variations of energy director designs (Courtesy Branson Ultrasonics Corp.)*

8.5.4   Shear Joints

Shear joints are typically used for applications that require a hermetic seal. It should be noted that hermetic seals can also be achieved with energy director joints, but the shear joint is usually preferred. Figure 8.16 shows a typical cross section of a shear joint along with recommended dimensions. Table 8.6 lists some of the advantages and disadvantages of shear joints.

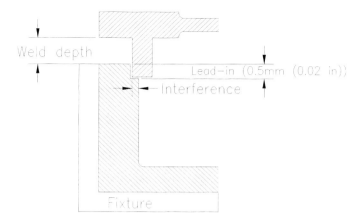

Figure 8.16 *Typical shear joint design*

It should be noted that one major disadvantages of the shear joint is the relatively high dimensional tolerance that is required to obtain a uniform weld. Therefore, when the part size increases, the shear joint is not generally recommended. Another disadvantage of the shear joint is that complex fixture design may be needed when part geometry cannot provide sufficient support for the shear joint. This is due to the fact that without providing support over the entire part, the part can deflect outwards as the upper part shears into the lower part. This will reduce the actual interference and result in poor welds. If the fixture provides uniform support over the entire weld surface, it is usually important that the fixture incorporates moving parts to allow removal of the assembled parts at the end of the weld cycle. One solution to this problem is to use a double shear joint, as seen in Fig. 8.17, which also shows other variations to shear joint design. Table 8.7 provides general guidelines for the dimensions and tolerances of shear joints for different part sizes.

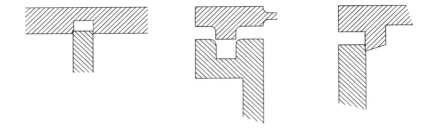

Figure 8.17 Other variations of shear joint designs

Table 8.6 Advantages and Disadvantages of the Shear Joint Design

Advantages	Disadvantages
High weld strength	Tight dimensional tolerance
Hermetic seal	Relatively high power dissipation
Work well with all materials	Complex fixtures may be required

Table 8.7 General Dimensional Guidelines for Shear Joints with Parts of Different Sizes

Maximum part dimension	Interference per side (range)	Part dimension Tolerance
Less than 18 mm	0.2 to 0.3 mm	±0.025 mm
18 to 35 mm	0.3 to 0.4 mm	±0.050 mm
Greater than 35 mm	0.4 to 0.5 mm	±0.075 mm

8.5.5   Part Design

Part design must provide for three major prerequisites crucial for successful welding:

- *Small initial contact.* A small initial contact area at the weld interface is required to assure that all the deformation (strain) occurs at the joint interface (faying surface) and to assure that the entire surface heats uniformly and quickly to the melt temperature of the thermoplastic material.

- *Assure proper alignment.* The part and sometimes the joint design should have built-in alignment to assure that the parts are positioned properly. Usually alignment is achieved through locating pins, part fit-up, or through joint design such as tongue and groove joints. Part alignment can be achieved by relying on the horn and fixture but this is not the preferred method.

- *Energy transfer.* The part should be designed to maximize the horn contact area to minimize induced locally high stresses at the horn/part interface, which also reduces the likelihood of part marking. In addition, the part should be designed to minimize the distance between the horn contact interface and joining surface. This distance should also remain constant throughout the part. Care should be taken to avoid voids that would prevent transmission of the ultrasonic vibration directly to the weld area (Fig. 8.18).

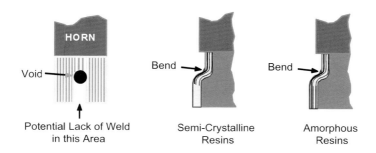

Figure 8.18 *Part design with good and poor transmission of ultrasonic vibration to weld interface (Courtesy Branson Ultrasonics Corp.)*

One of the first considerations given to part design is the weld distance (the distance between the horn/part interface and the weld interface). As described in Section 8.3.2, if this distance is less than 6 mm the weld is referred to as a "near-field" weld. If this distance is greater than 6 mm, the weld is referred to as a "far-field" weld. It is important to remember that when the weld distance is small compared to the wavelength in the material being welded, near-field ultrasonic welding applies. For near-field welding, it is often assumed that the amplitude of vibration at the energy director is similar to the amplitude of vibration applied by the horn. In contrast, in far-field welding wave propagation in the part must be considered when determining the amplitude of vibration that is experienced by the energy director at the weld interface. Whenever possible, it is best to design a part with a near-field design to minimize the risk of part marking, energy losses, part deflection, and machine over-loading condition. When it is not possible to design the parts for near-field welding, dimensions of the top part should be selected carefully to insure that the amplitude of vibration is maximized at the joint interface. As discussed by Menges and Potente [8], this usually means that the part length should be approximately half or full wavelength.

Another major consideration when designing a part is the possibility that a "sympathetic" mode of vibrations can be induced into the parts as a result of the ultrasonic vibrations. The mode can introduce damage to the part, when the part has a sympathetic mode of natural resonance close to or at the welding frequency or at a sub-harmonic frequency. Because it is very difficult to predict these modes even with FEA techniques [11], the following systematic approach will help solve these problems:

Diaphragming Effect (see Fig. 8.19)

Diaphragming occurs when a lid or a relatively flat and thin component is welded to another component. The undesired result of such a vibration is the heating and melting at the center of the lid. Some possible solutions are:

- Thicker wall section
- Internal support ribs
- Nodal plunger on horn
- Short weld time
- Higher or lower amplitudes
- Amplitude profiling ™[12]
- Using a higher frequency

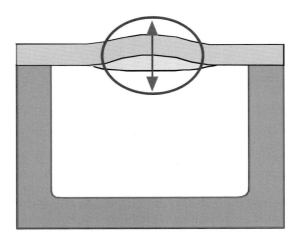

Figure 8.19 *Diaphragming effect*

Part Damage at Thin Walls, Fine Features, and/or Small Appendages (Fig. 8.20)

Under high amplitude of vibrations, these fine features vibrate and fracture can occur at their base. Possible solutions to problems associated with damage or fracture of fine appendages include:

- Addition of generous radius at the appendage base
- Dampening of the vibrations by external dampeners (foam rubber)
- Increase appendage thickness
- Use of a higher frequency

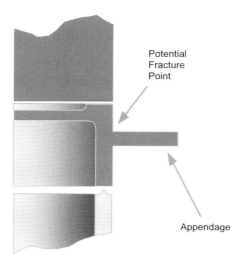

Figure 8.20 Potential vibration and fracture of appendages (Courtesy Branson Ultrasonics Corp.)

Sharp Corners Acting as Stress Concentrators (See Fig. 8.21)

Stress concentration at sharp corners can lead to localized heating during the welding process. Care should be taken during the design phase to make sure all radii are as generous as possible to minimize the risk of stress concentrations.

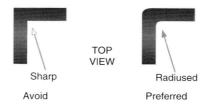

Figure 8.21 Sharp corners act as stress concentration points promoting local heating

Horn/Part Interface too Small

If the size of the horn/part interface is similar to that of the weld area, especially for semi-crystalline materials, the interface can heat during the weld cycle. This will result in part marking and damage. Figure 8.22 shows the importance of the horn/part interface when considering part design. If part marking is a problem, one possible solution is the use of a buffer sheet, i.e., a thin layer of consumable thermoplastics, usually PP or PE that is placed between the horn and the parts during each weld. The buffer sheet is designed to absorb the majority of the marking and is discarded after welding.

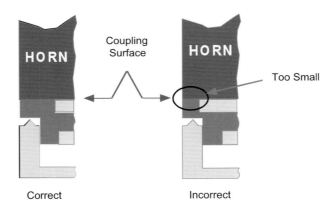

Figure 8.22 Importance of horn/part interface contact size (Courtesy Branson Ultrasonics Corp.)

8.5.6   Ultrasonic Insertion

Beyond standard welding techniques, it is possible to join parts that are not typically weldable, such as metals or wood products to thermoplastics by mechanical means with ultrasonic joining. One such technique is ultrasonic insertion. In this case, a metal insert, usually a threaded design, is "driven" into a thermoplastic substrate to accept a screw or similar device. Often, this is more economical than molded-in threads and it can have superior strength.

Part design for ultrasonic insertion will only be briefly reviewed because of the wide range of inserts that are available. Additional information can usually be obtained from both insert and ultrasonic welding equipment manufacturers. The most important consideration for ultrasonic insert part design is how to match the hole size to the profile of the insert. Most inserts require a lead-in and a simple tapered hole, see Fig. 8.23. The lead-in assures proper placement of the insert. Some inserts also require a hole with two different tapered profiles. While these types of inserts require slightly more complicated molds or complex drill bits, they tend to provide better stress distribution under load.

It is common to use inserts that have two knurled surfaces oriented in opposite directions to promote strength. When considering the size and shape of an insert, the most important criteria are: (1) pull-out strength and (2) torsional strength. Both of these are greatly dependent on the base material (material the insert is being driven into) and the ultrasonic insertion process parameters.

During the ultrasonic insertion process, the horn pushes a metal (usually brass) insert into the polymer. The metal-to-metal contact between the horn and the insert often results in excessive wear of the horn. To minimize horn wear, it is recommended to use a steel horn or to apply carbide or diamond coating to the face of the horn.

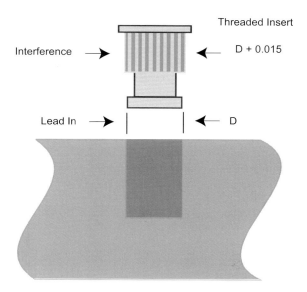

Figure 8.23 Part design for ultrasonic insertion (Courtesy Branson Ultrasonics Corp.)

8.5.7   Ultrasonic Staking

Ultrasonic staking is another common process that uses ultrasonic energy to join components. It is usually used to mechanically join dissimilar materials. This is accomplished by molding bosses on the thermoplastic part, which are inserted in holes in the second part. Ultrasonic heating and shaping is used to form heads (like a screw head) from these bosses. As shown in Fig. 8.24, there is a wide range of possible cross-sectional profiles for staking.

As might be expected, the horn must be designed to match the profile being used. It is common practice to use a horn with replaceable tips screwed into the front face. The main advantage of this design is to allow for relatively easy replacement of the tip if horn/tip wear is an issue. Tip wear can often be an issue when staking materials that are filled with glass or carbon fibers. Figure 8.25 shows examples of many of the commercially available tips.

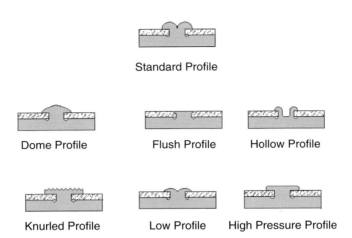

Standard Profile

Dome Profile   Flush Profile   Hollow Profile

Knurled Profile   Low Profile   High Pressure Profile

Figure 8.24 Cross-sectional views of different staking profiles (Courtesy Branson Ultrasonics Corp.)

Figure 8.25 Typical ultrasonic horn tips for staking (Courtesy Branson Ultrasonics Corp.)

8.5.8   Spot Welding

Ultrasonic spot welding is a technique commonly used to weld relatively thick thermo-plastic sheets (+ 4 mm). As the name implies, the resulting weld is located at a single spot. Because of this, the final seal is not continues and not hermetic. However, in many applica-tions, this type of joint is sufficient for the final application design. Spot welding has the advantage that often the equipment can be relatively simple: hand-held welding systems can be used in many of these applications. In addition, the applications do not require any joint design, allowing welds to be made at any location between two sheets.

The final joint configuration is defined by the spot welding tip. The tip design is usually based on the thickness of the top sheet to be welded. Figure 8.26 shows typical cross-sections of tip designs and the general dimensions based on top sheet thickness (T).

It is important to note that the final weld is limited to an area near the center of the spot weld. That is to say, very little adhesion is obtained away from the spot welding tip. This can be seen in Fig. 8.27, which illustrates a typical cross-section of spot weld. It is also important to note that there is no adhesion at the center of the spot weld, where the tip actually protrudes through the entire thickness of the top sheet.

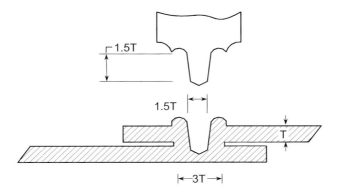

Figure 8.26 Cross-section of standard spot welding tip

Figure 8.27 Cross-section of typical spot weld (Courtesy Branson Ultrasonics Corp.)

8.6   Applications

It is beyond the scope of this chapter to review all applications that utilize ultrasonic welding for assembly, so only selected cases will be reviewed to give the reader an overview of the possible applications.

8.6.1   Plunge Welding Applications

Plunge welding is used extensively in the computer peripheral industry. Figure 8.28 shows photos of a variety of portable storage media, which are assembled using ultrasonic welding. One example is the 3.5 in. computer diskette, which is a very high volume product that requires an extremely fast assembly process. The outer housing of the diskette is assembled in less than 200 ms. In addition, a fabric liner is swaged in place in a separate ultrasonic welding process in less than 100 ms. Both of these operations are fully automated. Thus, because of the relatively low capital costs of the equipment and the high production rate, the process is cost effective.

Figure 8.28 *Computer storage media that are assembled using ultrasonic welding (Courtesy Branson Ultrasonics Corp.)*

Figure 8.29 shows a telephone hand set, another very popular application of ultrasonic welding, although this is not the only part on a telephone that is ultrasonically welded. For example, on mobile and cordless telephones, many of the batteries are assembled with ultrasonic welding. Again, one of the main reasons for selecting ultrasonic welding for this application is the production rate.

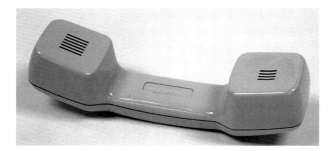

Figure 8.29 *Ultrasonically welded telephone handset (Courtesy Branson Ultrasonics Corp.)*

Another high-volume product assembled with ultrasonic welding is the common butane lighter (Fig. 8.30). This is an example of a critical application because the weld must be hermetically sealed and it must be able to withstand relatively high pressures and stresses. The main reason that ultrasonic welding was selected for this application is the ease of automation and the high production rates. It is important to note that a shear joint is used in this application and that there is no visible weld flash. This application is a prime example showing that with a proper joint design, fixture design, and machine setup even critical applications can utilize ultrasonic welding.

Figure 8.30 *Butane lighters are ultrasonically welded with no visible flash and a hermetic seal (Courtesy Branson Ultrasonics Corp.)*

Milk and juice containers are possible applications, requiring a hermetical seal with far lower pressure requirements (see Fig. 8.31). Sometimes only parts of these containers are sealed with ultrasonic welding. Ultrasonic welding can be applied because the paper is coated with PP and/or PE. Thus, when the material is sealed, only the outer plastic surface is actually being welded. Because of the limited amount of plastic material in the weld area, this application requires tight parameter settings. If the parts are over-welded, all the plastic film is displaced out of the weld zone and the seal will be weak. However, with the proper settings even the gable seal (top seal) can be sealed with ultrasonic welding although the thickness varies across the weld zone because of the cardboard folding processes.

Figure 8.31 *Milk and juice containers are often sealed with ultrasonic welding (Courtesy Branson Ultrasonics Corp.)*

Plunge welding is not restricted to rigid parts, e.g., in the production of facemasks from film or fabric material, the facemask is sealed around the edges and the retaining band is attached. Another very common non-rigid application welded with ultrasonics is blister packs (see Fig.8.32).

Figure 8.32 Example of film and fabric plunge ultrasonic welding applications (Courtesy Branson Ultrasonics Corp.)

Some applications benefit from the use of amplitude profiling™. With this technique the amplitude is modified/changed during the weld cycle to help control the squeeze flow of the molten polymer. In swaging applications, the flow of the material can be so fast that it shears out of the horn cavity resulting in excessive flash. With amplitude profiling, the process is slowed down, allowing more control of the heating and viscosity of the material. In Fig. 8.33, the benefits of amplitude profiling can be seen.

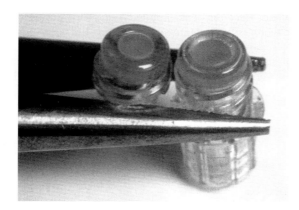

Figure 8.33 Superior welding results through Amplitude Profiling™ (Courtesy Branson Ultrasonics Corp.)

Ultrasonic insertion is not a true welding process since it usually involves insertion of a metallic component into the thermoplastic. There are many examples of ultrasonic insertion, the most common of which is the insertion of metal fasteners into automotive instrument panels. As shown in Fig. 8.34, it is not uncommon for one panel to have 10 to 20 fasteners inserted into it.

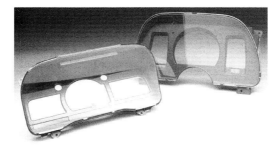

Figure 8.34 *Automotive instrument panels are held in place with metal inserts that are ultrasonically inserted-the mounting shoulders are seen in this figure (Courtesy Branson Ultrasonics Corp.)*

8.6.2   Continuous Applications

Continuous applications are not as common as plunge welded applications for ultrasonic welding. Probably the most common example of continuous sealing/welding for ultrasonic welding are feminine pads/napkins. As shown in Fig. 8.35, the napkins are cut and sealed simultaneously. In many similar applications, sealing of the cut edges prevents fraying.

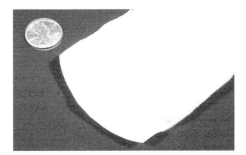

Figure 8.35 *Feminine pads are cut and sealed with ultrasonic energy (Courtesy Branson Ultrasonics Corp.)*

Another fabric that is often cut and sealed with ultrasonic welding, is nylon fabric for products such as belts and clothing articles (see Fig. 8.36). These applications are good examples in which the cut and seal geometry is complicated. These geometries can be created in continuous mode by using a rotating anvil with an electro-discharge machined pattern.

Another popular example of cut and seal applications is candy bar wrappers (see Fig. 8.37). Here the material is a film instead of a woven fabric. One of the major problems associated with this application is related to film thickness variations. The nominal thickness is 0.03 mm and even very small fluctuations in the material or machine setup can result in wide variations in weld quality.

Sealing and sewing of medical garments is another very important application of continuous ultrasonic welding (see Fig. 8.38). Prior to the introduction synthetic materials and ultrasonic welding, these cotton-based garments were sewed together which resulted in lose ends and fiber dust, both of which increased the risk of infection to a patient. With ultrasonic welding (sewing), the fiber ends are melted and sealed together, a process often compared to as cauterizing. This reduces dust production and minimizes the risk of infection to the patient.

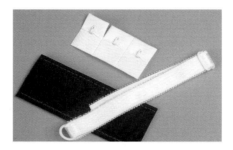

Figure 8.36 *Complex geometries can be cut and sealed with ultrasonic energy (Courtesy Branson Ultrasonics Corp.)*

Figure 8.37 *Many candy wrappers are sealed with ultrasonic welding (Courtesy Branson Ultrasonics Corp.)*

Figure 8.38 *Medical garments are ultrasonically sealed/sewed together to minimize fiber damage and dust (Courtesy Branson Ultrasonics Corp.)*

8.7   Material Weldability

8.7.1   Material Properties

Weldability is defined by the American Welding Society as "the capacity of a material to be welded under fabrication conditions imposed onto a specific suitably designed structure and to perform satisfactorily in the intended service." This broad definition can be applied to metals as well as polymers. For ultrasonic welding a thermoplastic's weldability can be measured in the following terms:

- *Energy Transfer.* This material property is directly proportional to a material's stiffness or storage modulus (defined in Section 8.3.1) at the operating frequency and it defines how well the material transmits sound. Materials with low storage modulus, such as elastomers, PP, and PE, do not transfer sound well and thus tend to be difficult to weld unless the effects of this property are minimized by part design. For example, when welding low storage modulus materials, the distance between the horn and weld zone should be minimized (near field welding).

- *Energy dissipation.* This property defines how well a material can transform mechanical energy (sound waves) into thermal energy (heat). This property is directly related to the material's loss modulus (discussed in Section 8.3.1). The higher a material's loss modulus, the better the material can convert the applied sound energy into heat. Materials with a relatively high loss modulus, such as PS can easily be heated with ultrasonic energy.

- *Self-adhesion or healing.* This property defines how well a material will heal to itself at the interface once in a molten state. That is to say, it is a measure of how quickly and to what extent will the molten interface bond to itself. The process of self-healing can be divided into two major stages; self-wetting, where the melted polymer wets melted polymer from the second part, and intermolecular diffusion, where polymer molecules diffuse across the interface and get entangled. As described in Section 8.3.1, these are the stages of squeeze flow and intermolecular diffusion. In the case of dissimilar polymers, welding or healing is still possible if the polymers will readily wet each other and if they are miscible in each other. It recommended that in the case of welding different polymers, they have similar viscosities and similar melting temperatures. It is important to remember that the melt viscosity will affect the kinetics of wetting and diffusion. Therefore, for high molecular weight polymers, such as many fluoropolymers and ultra-high molecular weight polyethylene, wetting and intermolecular diffusion takes so long that they become difficult if not impossible to weld.

In summary, for a material to be considered easily weldable with ultrasonics, the material should be able to transfer sound energy from the horn/part interface to the weld zone and then be able to easily convert the mechanical energy into thermal energy. In many cases it is not easy to find a material that has good energy transfer and dissipation properties. For example, PP has excellent dissipation properties but very poor transfer properties, while PC displays quite the opposite behavior with excellent transfer properties but relatively poor dissipation properties.

8.7.2   Fillers

Fillers, such as glass or mineral powders or mold release agents, can greatly affect a material's weldability. Fillers affect the weldability of each plastic differently and understanding the neat resins weldability is important to understand how fillers will influence this property. For example, adding small amounts (1 to 10%) of short glass fibers (SGF) to PP and PE can increase their energy transfer and thus increase their weldability. Adding more than 15 to 20% SGF to PP and PE can decrease their weldability because the squeeze flow during welding results in adverse orientation of the fibers and increased concentration of the fibers at the bond line, thereby reducing the weld strength.. Thus, as a general guideline, adding small amounts of stiffening fillers, such as SGF or mineral fillers (talc), to low modulus materials increases their ability to transfer energy and thus their weldability. However, adding more than 10 to 15% of most fillers, to most thermoplastics reduces their weldability. Table 8.8 provides some general guidelines of the effect of a few common fillers on the weldability on some common polymers. It should be noted that these are very general guidelines and analysis of the effect of specific fillers on the weldability of a specific polymer can only be done through experimental testing.

Table 8.8 *General Guidelines for the Effect of Fillers on the Ultrasonic Weldability of Selected Thermoplastics*

Material	SGF		Mineral		Internal Mold Release	
	(0 to 10%)	(+15)	(0 to 10%)	(+15)	(0 to 3%)	(+3)
PC	No effect	Decreases	No effect	Decreases	Decreases	Decreases
ABS	No effect	Decreases	No effect	Decreases	Decreases	Decreases
PS	No effect	Decreases	No effect	Decreases	Decreases	Decreases
PP	Increases	Decreases	Increases	Decreases	Decreases	Decreases
PE	Increases	Decreases	Increases	Decreases	Decreases	Decreases
PA	Increases	Decreases	Increases	Decreases	Decreases	Decreases
ABS/PC	No effect	Decreases	No effect	Decreases	Decreases	Decreases

8.7.3   Materials

In order to accurately determine materials weldability to itself or to other materials, a weldability study must be completed. The newly published AWS G1.2M [13] standard provides guidelines for completing such a test. Figure 8.39 provides qualitative estimation of the ultrasonic weldability of common thermoplastics and general weldability between dissimilar materials.

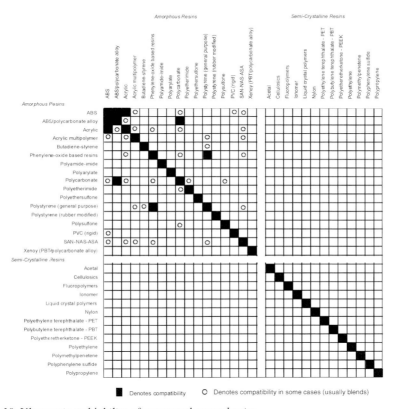

Figure 8.39 *Ultrasonic weldability of common thermoplastics*

Table 8.9, provides additional information on selected materials and their amenability for ultrasonic joining (welding, staking, and insertion).

Because film and fabric type applications are significantly different compared to rigid material applications Table 8.10 is provided for general guidelines on material weldability for film and fabric applications.

Table 8.9 *Reference Guide for Ease of Joining with Ultrasonics for Thermoplastics (Courtesy Brarson Ultrasonics Corp.)*

Material	Welding		Swaging/ Staking	Insertion	Spot Welding
	Near*	Far*			
Amorphous Polymers					
ABS	1	2	1	1	1
ABS/polycarbonate alloy	2	2	2	1	1
Acrylic	2	3	3	1	1
Butadiene-styrene	2	3	2	2	2
Phenylene-oxide based resins	2	2	2	1	1
Polycarbonate (a)	2	2	3	2	2
Polyetherimide	2	4	4	3	3
Polyethersulfone (a)	2	4	4	4	4
Polystyrene (general purpose)	1	1	4	2	3
Polystyrene (rubber modified)	2	2	1	1	1
Polysulfone (a)	2	3	3	2	3
PVC (rigid)	3	4	2	1	3
SAN-NAS-ASA	1	1	3	2	3
PBT/polycarbonate alloy	2	4	3	2	2
Semi-Crystalline Polymers (b)					
Acetal	2	3	3	2	2
Cellulosics	3	5	2	1	3
Fluoropolymers	5	5	5	5	5
Liquid crystal polymers (c)	3	4	4	4	3
Nylon (a)	2	4	3	2	2
Polyester, thermoplastic					
Polyethylene terephthalate/PET	3	4	4	3	3
Polybutylene terephthalate/PBT	3	4	4	3	3
Polyetheretherketone – PEEK (c)	3	4	4	3	3
Polyethylene	4	5	2	1	2
Polymethylpentene	4	5	2	1	2
Polyphenylene sulfide	3	4	4	2	3
Polypropylene	3	4	2	1	2

Code: 1 = Easiest, 5 = difficult

Table 8.10 *General Weldability Guidelines for Film and Fabric Materials*

Material	Woven	Non-woven	Knitted	Coated Materials	Laminates	Films
Acylic	4	Na	4	Na	Na	Tacked only
EVA	Na	Na	Na	2	Na	1
Nylon	2	2	2	2	2	2 (cut & seal only)
Polyester	2	1	2	1	1	1
Polyethylene	Na	1	Na	1	1	4–5 (+ 0.076 mm)
Polypropylene	1	1	2	1	1	1
PVC	3-5	Na	Na	3-5	3-5	3–5 (grade dependent)
Saran	Na	Na	Na	1	Na	1 (cut & seal only)
Surlyn	Na	Na	Na	1	Na	NA
Urethane	Na	Na	Na	1	Na	1 (thermoplastics only)
Natural fibers with fusibles	2	2	2	2	2	Na

Code:1 = Easiest; 5 = Most Difficult; Na = generally not seen in this form

8.8   Setup and Troubleshooting

8.8.1   Basic Procedures for Ultrasonic Welder Setup

Initial equipment setup can be divided into three primary procedures:

1) Conditioning of converter/booster/horn stack interface and resonance
2) Alignment and leveling of ultrasonic horn/part/fixture as a system
3) Selecting optimal process parameters

Proper setup of each of these steps can be critical to successful ultrasonic welding, and are detailed in the following sections.

8.8.1.1   Conditioning of Converter/Booster/Horn Stack (Stack Assembly)

Intimate contact between the converter and booster and the booster and horn is critical for proper operation. Improper stack assembly can result in high noise levels, poor tuning (will require higher power consumption during idle test tuning trials), and wasted power. If the equipment is operated continuously, this may eventually lead to heat damage to the converter and potential overloading of the power supply. The following steps are recommended for proper preparation of the stack:

- Examine interfaces carefully to inspect for damage or corrosion. Wipe interfaces with clean cloth or paper towel. If the interfaces are not damaged or corroded, apply a small amount of gel liquid such as silicon gel to improve ultrasonic wave propagation. Use an amount about the size of a paper match head.
- Before inserting a stud to the threaded hole for assembly, clean the threaded hole and stud with a clean file or wire brush, and cloth/towel.
- When tightening the studs, refer to industry standards for torque to avoid under- or over-tightening. If this information is not available from the equipment manufacturer, it is recommended to use 33 Nm of torque for 3/8–24 studs and 47 Nm of torque for ½–20 studs.
- After successful completion of previous steps, the stack assembly is placed into the equipment and is evaluated under in-air tuning conditions. If loud noise is emitted and/ or a high level of power (20% or more) is observed, repeat steps 1 through 3 to determine the source of the problem.

8.8.1.2   Alignment and Leveling of Ultrasonic Horn/Part/Fixture as a System

Maintaining proper leveling and alignment between the horn/stack assembly and the parts is critical to making uniform welds. Poor alignment, either horizontally or vertically, can result in over-welding at one side while under-welding at the other side. The following procedures are recommended for achieving uniform contact:

- Slightly loosen the bolts on the actuator door such that the horn can be rotated and aligned with the part. Also, loosen the bolts on the fixture to enable alignment.
- Adjust actuator height to allow enough space to fit the part into the fixture. Most machines have toggles at the sides to adjust the actuator height. Check that the height does not exceed the stroke length of the actuator cylinder.
- Insert the part in the fixture.
- Either turn off the power switch or set air pressure to zero to lower the horn manually. Then align horn, part, and fixture.
- Tighten the bolts on the actuator door and fixture.
- Use the so called "carbon paper" method as depicted in Fig. 8.40 to evaluate the alignment and leveling of the ultrasonic horn/part/fixture. After proper setting and tightening of the bolts, use the horn-down switch (if available) to generate an imprint of the carbon paper image on white paper. If a horn-down switch is not available, reduce

the air pressure to zero and lower the horn manually to form the carbon paper imprint. Based on the uniformity of the imprint, adjust the fixture alignment accordingly until a uniform carbon imprint is obtained.

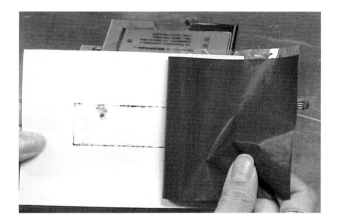

Figure 8.40 Carbon paper method for testing fixture/part/horn alignment

8.8.1.3   Selecting the Optimal Process Parameters

The main ultrasonic welding process parameters are trigger force, down speed, weld time, amplitude of vibration, welding pressure, hold time, and hold pressure. Most modern ultrasonic welding machines with upgrade options also provide Statistical Process Control (SPC) capability for better welding process control. With SPC, peak power, energy, melt down distance, and absolute distance during the welding process can be controlled. Depending on material characteristics, part geometry, and size of the parts, any of the SPC parameters can be used to monitor the process. While there are many parameters that need to be set for optimal welding, vibration amplitude of has the most influence on weld quality. Table 8.11 provides the typical range of peak-to-peak vibration amplitudes that should be used at 20 kHz for various materials and applications. For alternative frequencies, multiply the amplitudes by a factor that is equal to the alternative frequency/20 kHz. For example, if the recommended 20 kHz amplitude is 50 µm, and the equipment is actually 40 kHz, then the recommended starting amplitude is 25 µm.

In many applications, design of experiment (DOE) methods can be used to quickly and efficiently find the near-optimum welding conditions. DOE is based on different statistical methods such as central composites, box-Behnken, simplex centroid/lattice, and others. Prior to application of DOE, it is necessary to perform screening experiments to determine the range for the various process parameters. Following a specific DOE method, a set of experiments is performed. The test results from these experiments are then used together with statistical regression techniques to determine the optimal welding conditions. It is often recommended to test the final process parameters by welding many samples under these conditions to determine the robustness of the process. Below are the recommended steps for performing DOE for the process optimization:

Table 8.11 *Typical Range of Vibration Amplitudes at 20 kHz for Various Materials and Applications*

Material	Welding (μm)	Insertion (μm)	Staking (μm)
PS	20–40	20–40	20–30
PC	50–70	50–70	40–60
ABS	40–60	40–60	30–50
PP	70–90	70–90	50–70
PE	70–90	70–90	50–70
PA	60–80	60–80	60–80
PPO	50–90	50–80	40–70
PEI	60–125	60–100	40–100
PVC	40–75	40–75	30–60
SAN/NAS	30–65	30–65	30–60
Acetal	75–125	75–100	50–100
PEEK	60–125	60–100	60–100
PPS	80–125	75–100	75–100
PI	60–125	60–100	60–100
PBT	60–125	60–100	60–100

1) Determine noise factors such as usage of regrind of material, different versions and kinds of materials.

2) Select control factors which will affect the weld quality such as weld time, weld pressures, hold time, etc.

3) Determine response parameters based on the focus of the DOE. Acceptable responses could be tensile strength, marking on surface, peak power, melt down distance, energy, etc.

4) Upon completion of DOE, depending on the noise factors, control factors, and responses, different regression methods can be used to analyze the results and determine the optimal process conditions for the regression parameters.

5) Testing of the optimum conditions is highly recommended to determine how robust the process is. This can be done by testing many samples using the optimum or near optimum welding parameters and evaluating the weld quality to insure that it is within an acceptable range. Otherwise, steps 2–4 must be repeated until robust welding conditions are established.

8.8.2   Troubleshooting Guide

The following guidelines for troubleshooting will summarize the appropriate diagnostics of common problems that may be encountered while performing the ultrasonic assembly processes. These recommendations are for altering primary welding parameters such as weld time, weld pressure, hold time, hold pressure, and amplitude of vibration to circumvent the troubles. Since ultrasonic welding machines can be utilized for welding, staking, and insertion, recommendations for these are summarized in separate tables (Tables 8.12 – 8.15).

Table 8.12 *Troubleshooting Guidelines for Ultrasonic Welding*

Problem	Probable Causes	Recommendations
Over-welding	Too much energy input for welding	Reduce weld pressure, weld time, energy, welding distance
		Reduce amplitude of vibration by using lower power setup and/or lower gain booster
		Reduce down speed
Under-welding	Insufficient energy input for welding	Do the counter actions described for over-welding problem
	Energy loss to fixture (if fixture is made of urethane)	Change to a fixture made of more rigid material
Inconsistent welding from part-to-part	Part dimension variation	Run statistical study if a pattern develops with certain cavity combinations
		Check part tolerances/dimensions, molding conditions, and cavity dimension variation
	Mold release	Clean mating surface or replace mold release
	Variation in material characteristics	Check regrind and/or filler content variation
		Check molding condition variation
		Check moisture content
	Variation in utility	Check line voltage and air pressure fluctuation

Table 8.12 *Troubleshooting Guidelines for Ultrasonic Welding (Continuation)*

Problem	Probable Causes	Recommendations
Marking	Horn heats up	Decrease amplitude
		Check for cracked horn, booster, or converter
		Check for loosened stud and horn
		Utilize air nozzles for cooling
	Weld cycle is too long	Increase amplitude and/or pressure
		Adjust dynamic air pressure
	Improper fit of part to fixture	Check for proper fixture support
		Redesign fixture
Flash	Energy director is too large	Reduce size of energy director, weld time, pressure
	Over-welding	Refer to recommendation in over-welding
	Shear interference is too great	Reduce amount of interference
	Insufficient flash trap	Add flash trap
		Increase flash trap

Table 8.13 *Troubleshooting Guidelines for Ultrasonic Insertion*

Problems	Probable causes	Recommendation
Inserts pull out easily	Insufficient interference between hole and insert	Reduce hole size
	Inserts get pushed in boss before plastic melts	Increase amplitude and/or reduce the weld pressure
		Use pre-trigger on
	Horn retracts before the plastic around the inserts is solidified	Increase hold time

Problems	Probable causes	Recommendation
Inconsistent insertion	The plastic is not melting consistently around all inserts	Increase amplitude
	Inserts are seated at different height	If using a process controller, weld under specified distance (absolute or collapse)
		Use mechanical stop, if applicable
The plastic around boss cracks	The insert is pushed in before the plastic gets melted	Reduce the down speed, weld pressure, and/or amplitude
		Set pre-trigger on
		Increase the thickness of the boss wall

Table 8.14 Troubleshooting Guidelines for Ultrasonic Staking

Problems	Probable causes	Recommendations
Stake head is not uniform	Staking cavity at horn is too large or insufficient volume in the stud	Reduce cavity size or increase the stud height or diameter
Excessive flash is formed around the stake head	Staking cavity is too small or stud volume is too large	Increase staking cavity or reduce the stud height and/or diameter
Parts are loose after staking	The hole diameter is too large with respect to stud diameter	Reduce hole diameter
	The holding force was released before the molten stud was solidified	Increase hold time
		Increase the stud diameter
		During holding period, external clamps or nodal plunger can be used
Premature staking head is formed at the end of cycle	Insufficient weld time or energy input	Increase weld time/amplitude, and energy

Table 8.14   *Troubleshooting Guidelines for Ultrasonic Staking (Continuation)*

Problems	Probable causes	Recommendations
Severe marking and distortion on opposite side of stake head	Inappropriate fixture alignment	Improve the fixture alignment
	Clamping pressure is too high	Reduce clamp pressure. If heat damages surface, placing a metal foundation at the fixture surface will serve as a heat sink to reduce marking
	Amplitude or energy input is too high	Reduce amplitude and/or energy input
The stud is broken at its base	There is stress concentration caused by a sharp corner near the base of the stud	Introduce radius at the stud corner
	Too much pressure is applied before melting.	Set up pre-trigger on
	The stud is aligned at the center of horn	Align the stud at the center of horn cavity

Table 8.15 *Troubleshooting Guidelines for Ultrasonic Equipment/Setup*

Problems	Probable causes	Recommendations
Heat develops at the interface between horn/ booster or booster/trans- ducer	Loose connection at the interface	Disassemble the stack and clean the interface before retighten the stack by the manufacturer's suggested torque.
	A mounting stud is cracked	Replace stud.
	A stud is loosened	Remove loose stud. Inspect the stud and retighten the stud.
	There is a dirty interface between horn/booster or transducer/booster	Clean interfaces with clean cloth or paper towel and apply a thin film of grease before re- assembly.
Noise comes from the stack	A loosened interface between components	Check the stack assembly and retighten with the recommended torque
	The stack component gets cracked	Replace any component as needed
	The stack frequency is out of the resonance range	Retune generator

Welder will not tune in air	Defective horn, converter, booster, stud, or junction	Replace stack components and junctions as needed
	Defective power supplier	Replace power supplier
	Improper horn	Check resonance frequency of horn and compare it with weld frequency range. If needed, replace it
Welder tunes in air but stalls at welding	Defective horn, converter, booster, stud, or junction	Replace stack components and junctions as needed
	Defective power supplier	Replace power supplier
	Unstable horn	Redesign the horn
	Horn requires high starting energy	Use pretrigger, reduce pretrigger force, use the slower pressure buildup rate, and/or use lower amplitude gain horn

More detailed examples and recommendation can be found in many handbooks that are provided by ultrasonic welding machine suppliers and in the upcoming American Welding Society Guidelines for Ultrasonic Assembly.

8.9   References

1. Ultrasonic Plastics Assembly, Branson Ultrasonics Inc., Danbury, CT 1979

2. Ensminger, D., Ultrasonics: Fundamentals, Technology, Applications, 2nd Edition, Marcel Dekker, Inc., New York, 1988

3. Benatar, A. and Gutowski, T.G., "Ultrasonic Welding of PEEK Graphite APC-2 Composites." *Polymer Engineering and Science*, **29**, 1705, 1989

4   Read, B.E. and Dean, G.D., The Determination of the Dynamic Properties of Polymers and Composites, John Wiley and Sons, Inc., New York, 1978

5. Ferry, J.D., Viscoelastic Properties of Polymers, Chapter 7, John Wiley and Sons, Inc., New York, 1970

6. DeGennes, P.G., "Reptation of a Polymer Chain in the Presence of Fixed Obstacles," *Journal of Chemical Physics*, **55**, 572, 1971

7. DeGennes, P.G., "Entangled Polymers," *Physics Today*, June 1983, p. 33

8. Menges, G. and Potente, H., "Studies on the Weldability of Thermoplastic Materials by Ultrasound," *Proc. of Welding in the World*, **9**, 1971

9. Benatar, A. and Cheng, Z., "Ultrasonic Welding of Thermoplastics in the Far-Field," *Polymer Engineering and Science*, **29**, 1699, 1989

10. Grewell, D., "Amplitude Profiling; A Study in Ultrasonic Welding of Plastics," Proceedings of the 54th Annual Technical Conference, Society of Plastics Engineers, Brookfield, CT, May 1996

11. Ng, W.C. and Benatar, A., "Analysis of the Vibration and Viscoelastic Heating of Polystyrene Beams," Proceedings of the 54th Annual Technical Conference, Society of Plastics Engineers, Brookfield, CT, May 1996

12. Grewell, D., "Amplitude and Force Profiling in Ultrasonic Welding of Thermoplastics," Proceedings of the 55th Annual Technical Conference, Society of Plastics Engineers, Brookfield, CT, May 1997

13. AWS G1.2M/G1.2:1999, "Specifications for Standardized Ultrasonic Welding Tests Specimen for Thermoplastics," American Welding Society, Miami, Florida, 1999

9   Linear and Orbital Vibration Welding

Ian Froment

9.1   Introduction

Friction welding is a widely used technique for the joining of thermoplastic materials in industrial applications [1, 2, 3]. The thermoplastic friction welding processes simply involves the rubbing together of two thermoplastic interfaces until the surfaces become molten and flow together to produce a weld. Friction welding, like other thermoplastic welding processes, is used when the complexity of the injection molded component prevents its manufacture as a one-piece-component. The major user of the friction welding processes is the automotive industry for mass production of welded components such as car bumpers, engine air intake manifolds, and other under-the-hood components.

There are four separately identifiable thermoplastic friction welding processes:

- Linear vibration,
- Orbital vibration,
- Angular vibration and
- Spin or rotational friction welding.

Linear vibration welding is the most common used friction welding process for thermoplastics. It uses a linear reciprocating motion at the joint interface to produce the heat to produce the weld. In orbital vibration welding an orbital motion that produces a constant velocity at the joint interface is used to produce the necessary heating for the welding operation. Both the linear vibration and orbital vibration processes are often used in applications that are considered too large for the ultrasonic welding process. Angular friction welding is designed to weld circular geometry components where final alignment of the components is critical to the application. In this process, an angular reciprocating motion is used to produce the required friction between the two components being welded. Since the advent of microprocessor control, the angular friction welding process has been all but superseded by spin welding equipment with rotational positional control and has therefore disappeared from industrial applications. Spin welding and angular friction welding are detailed in Chapter 10.

The vibration welding processes have a number of advantages when compared to other thermoplastic welding processes. The welding times are relatively short for large component geometry when compared to processes such as hot plate welding. For example, a simple, four sided box measuring 200 mm x 200 mm with a wall thickness of 3 mm, would typically have a vibration welding machine cycle time of 50 to 60 s (part to part). An equivalent hot plate welded component would have a machine cycle time in the order of 120 to 180 s.

On the other hand, the capital cost of the equipment, particularly linear vibration, is high compared other thermoplastic welding processes. A linear vibration-welding machine may cost between five and six times as much as an ultrasonic welding machine [4]. However, when comparing the costs of vibration welding to other processes, it is important to note that a single vibration machine can join as much weld area as several ultrasonic welding machines. Thus, the costs may actually be comparable. In addition to machine costs, tooling/fixtures for vibration welding can also be relatively expensive. In general, the vibration process can produce around four times the number of components that can be manufactured using the hot plate welding process. Compared to ultrasonic welding the machine cycles are approximately four to five time longer although geometry of the component will become a limiting factor in the ultrasonic welding process.

Thin wall (typically less than 1.5 mm) applications can pose difficulties for the linear friction welding process, particularly when the direction of the vibration is across the wall of the component. Careful component design is required to prevent wall flexing in these applications. A more suitable process for these applications is the orbital vibration welding process, where each part of the component wall will experience both the longitudinal and transverse motion and constant interfacial velocity [4].

Finally, the need to have a single plane joint in the direction of the vibration is a distinct disadvantage when designing a component for the vibration (linear and orbital) welding processes. A slight incline in the joint of about 10° can be tolerated in the linear vibration welding components, although it is recommended to avoid such features when designing components for vibration welding.

9.2   Process Description

9.2.1   Linear Vibration Welding

Linear vibration friction welding of thermoplastics involves the rubbing together under axial force of two injection molded thermoplastic components in a linear, sinusoidal recip-rocating motion. The relative motion is illustrated in Fig. 9.1. The frictional heat generated at the weld interface by the rubbing action must be sufficient to melt and allow the flow of thermoplastic material. A combination of weld time, weld forces (pressure at the interface) and interfacial velocity generates the heat required to melt the weld interface. The interfa-cial velocity is a function of the amplitude (A) and the machine vibration frequency (ω) (see Section 9.3).

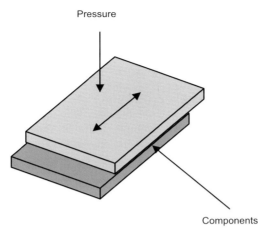

Pressure

Components

Figure 9.1 Motion produced by linear friction welding

Table 9.1 provides an overview over the linear vibration welding cycle along with typical times for each step in the cycle.

Table 9.1 Typical Linear Vibration Welding Cycle

Step	Typical time
Load components	Manually (3 to 5 s)/automated (1 to 3 s)
Activate cycle (palm buttons or automated)	Manually (1 to 2 s)/automated (0 to 0.5 s)
Door closes	0.5 to 2 s
Lower tooling platen raises	1 to 4 s
Vibration weld cycle	8 to 15 s
Hold cycle	5 to 15 s
Lower tooling platen lowers	1 to 4 s
Door opens	0.5 to 2 s
Remove parts	Manually (3 to 5 s)/automated (1 to 3 s)

The vibration welding machine cycle commences when the parts are placed into the tooling. The parts can be either pre-assembled and placed into the lower tooling platen or one component-half placed in the lower tooling platen and the other component-half placed in the upper. Components placed in the upper tooling can be held in place using a vacuum system or another mechanical clamping design. Pressing two palm buttons activates the manual weld cycle. Two buttons are employed to ensure that the operator's hands are simultaneously occupied to prevent entrapment in the moving machine parts. A door is then closed, again to prevent entrapment during welding, and to seal the noise reduction cabinet. Finally, the lower tooling platen raises and the actual welding cycle begins with the vibration and cooling stages. The vibration and cooling cycle can take up to 15 s each, depending on the size and geometry of the component being welded. The frequency of the vibration ranges from 100 to 300 Hz, depending on the type of machine employed. In a typical 10 s weld cycle, the parts will be rubbed together between 1000 and 3000 times. The welding machine will take between two and four cycles of vibration to achieve the maximum amplitude and overcome inertia. Because this is only a fraction of the vibration cycle, this does not have an effect on the quality of the welded component.

The welding machine cycle is completed when the tooling platen lowers, with the welded component in position, and the doors open. The components can then either be removed manually or by an appropriate component handling system.

9.2.2   Orbital Vibration Welding

Orbital friction welding involves the rubbing together under axial force of injection molded components in an orbital motion. The motion at the interface between the two components is effectively small circular or elliptical with a diameter of approximately 1 to 2 mm. The frictional heat generated at the weld interface by the rubbing action must be sufficient to melt and allow the flow of thermoplastic material at the interface. A combination of weld time, weld force (pressure at the interface), and interfacial velocity generates sufficient heat to melt the weld interface. The interfacial velocity is a function of the amplitude and the machine vibration frequency (see Section 9.3). This technique is more suitable for components that have relatively thin walls (< 2 mm), where linear vibration welding would not normally work. Unlike linear vibration welding, the relative motion of the two parts at the weld interface is the same at all points on the component perimeter. In linear vibration welding there are two types of relative motion; one transverse to the component wall and one longitudinal to the component wall. A square box component will experience the longitudinal motion along two sides whilst the other two sides will experience the transverse motion. In orbital welding, each point of the welding interface will experience both the transverse and longitudinal motion depending upon direction of the orbit compared to component wall at a given time. This can be approximated by longitudinal movement during two quarters of the orbit and transverse movement during the other two quarters of the orbit.

As with the linear vibration welding process, there are a number of steps to the orbital vibration welding machine cycle. Many of these steps require a similar time to the linear vibration process shown in Table 9.2 along with typical times for each step in the cycle.

The orbital vibration welding process tends to be used for smaller size applications than the linear vibration process, hence the vibration and cooling times are generally shorter. As with the linear vibration welding process, the vibration machine cycle commences when the components are loaded into the welding machine. Components are loaded into the orbital vibration-welding machine in a similar manner (see Section 9.2.1) and the machine cycle is activated by two palm buttons or automatic control system. Again, the machine cabinet is insulated to reduce noise during the welding cycle.

Table 9.2 *Typical Orbital Vibration Welding Cycle*

Step	Typical time
Load components	Manually (3 to 5 s)/automated (1 to 3 s)
Activate cycle (palm buttons or automated)	Manually (1 to 2 s)/automated (0 to 0.5 s)
Door closes	0.5 to 2 s
Lower tooling platen raises	1 to 4 s
Vibration weld cycle	4 to 10 s
Hold cycle	5 to 10 s
Lower tooling platen lowers	1 to 4 s
Door opens	0.5 to 2 s
Remove parts	Manually (3 to 5 s)/automated (1 to 3 s)

9.2.3   Process Control

Process control in vibration welding is similar for both the orbital and linear vibration welding processes. In vibration welding, for example, there are many modes of operation/control, including welding by time or displacement, and the particular mode will vary depending upon the equipment employed, manufacturer of the equipment, and application. With the advent of the microprocessor, equipment is controlled by programmable logic control allowing feedback of both the time and displacement welding parameters simultaneously.

In the first, most common, method/mode, welding by time, the parts are loaded into the vibration-welding machine and the machine cycle is initiated. When the two parts come into contact, the microprocessor (controller) senses that the lower tooling platen has stopped traveling upwards and therefore the vibration cycle can be started. There is usually a short time after contact (typically 1 s) to allow the force to stabilize prior to vibration. When welding by time, the vibrations are engaged for a fixed, pre-selected length of time and then stopped, allowing the cooling cycle to begin. During the vibration welding cycle, the microprocessor can continuously monitor the displacement of the lower tooling platen. This is achieved using a displacement transducer between the moving lower tooling platen and the vertically fixed vibration head. The displacement is monitored to give a degree of quality control to the welding operation. Prior to starting the vibration welding process, displacement limits can be set within the machine. If during a given welding operation, the lower limit is not achieved or the upper limit is exceeded, the machine will produce an alarm for the operator to inspect the component.

When welding by displacement, the microprocessor (controller) monitors the displacement of the lower tooling platen during welding and discontinues the vibration-welding head when the desired level of material displacement is achieved. The displacement transducer will typically be a linear encoder, ensuring good accuracy of measurement while the component is collapsing/displacing. As with welding by time, the lower tooling platen will rise until the component parts contact and a short period of time will elapse prior to switching on the vibrations to allow the force to stabilize. In order to assure a high degree of quality, welding time limits can be set in the machine. If the welding operation fails to achieve the lower time limit or it exceeds the upper time limit, the operator can remove the component from the production line and inspect it.

9.3   Process Physics

Both vibration (linear and orbital) welding processes function by producing heat at the welding interface due the rubbing together of the two surfaces being welded. The melting process occurs in four discrete phases [5, 6, 7]:

1. Initial heating of the thermoplastic resin through the friction of solid-to-solid interfacial heating (Phase I)

2. Intermittent heating of the thermoplastic resin through shear within a thin layer of molten thermoplastic material, transitional phase between Phase I and Phase III (Phase II)

3. Stationary thermoplastic temperature, steady state melt displacement (Phase III)

4. Cooling down of the molten thermoplastic (Phase IV)

These phases are shown graphically in Fig. 9.2. In the initial solid phase of the friction cycle, the weld area is heated up by frictional energy until the interfaces reach the melting temperature of the thermoplastic. The fundamental approach to modeling heat generation is based on the simple equation relating force and velocity to power (P), see Eq. 9.1.

$$\text{Power} = \text{Force} \times \text{Velocity} \tag{9.1}$$

This equation shows that power is directly proportional to velocity (v) and force (F_f). This is important to note because the velocity at the weld interface is defined in Eq. 9.2 for both linear and orbital welding. In most orbital application the motion is circular where the amplitude in both the X and Y direction (A_x, A_y) are the same, thus this will be consider the primary motion. If the amplitude in both directions is not equal resulting in an elliptical motion, Eq. 9.2 will need to be modified accordingly.

The relevant force for frictional heat is the frictional force produced when two solid-to-solid interfaces are slid past one another, as seen in Fig. 9.3.

The frictional force is defined in Eq. 9.3 [8], where F_n is the normal force applied to the parts and μ is the coefficient of friction between the two interfaces. Since friction welding is a process of motion, only the dynamic coefficient friction is important to modeling the process. In general, the dynamic coefficient of plastics can vary greatly. For example, an interface of PC on PC can have a dynamic coefficient as high as 0.45 while the Teflon/ Teflon® coefficient is 0.04. Table 9.3 lists the coefficient of frictions for selected plastics.

LINEAR :

$$v = \frac{du}{dt} = A\omega \cdot \cos(\omega t)\ldots \text{ over multiple cycles the RMS velocity } \frac{A\omega}{\sqrt{2}}$$

ORBITAL :

$$v_x = \frac{du_x}{dt} = A_x \, \omega \cdot \cos(\omega t)$$

$$v_{yx} = \frac{du_y}{dt} = A_y\omega \cdot \sin(\omega t)$$

IF $A_x = A_y = A$ (Circle)

$$v = \sqrt{v_x^2 + v_y^2} = A\omega \tag{9.2}$$

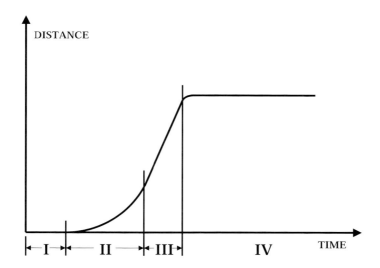

Figure 9.2 *Schematic of displacement vs. time for linear and orbital vibration welding*

$$F_f = F_n \cdot \mu \tag{9.3}$$

Because some materials have a very low coefficient of friction, such as Teflon®, they are not well suited to friction welding because the amount of heat generated through frictional heating is not sufficient.

With the velocity and forces defined, it is possible to define the average power dissipation generated during Phase I (solid/solid) as seen in Eq 9.4.

LINEAR

$$P = Fxv = \ldots \text{over multiple cycles the RMS power dissipation is } \frac{2\mu FA\omega}{\pi}$$

ORBITAL

$$P = Fxv = \mu FA\omega \tag{9.4}$$

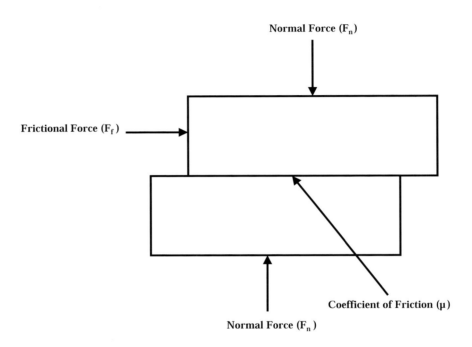

Figure 9.3 Example of frictional force

Table 9.3 Typical Values of Coefficient of Friction for Selected Thermoplastics [9]

Material	Coefficient of friction (μ)
Nylon	0.20 – 0.40
Acetal	0.15 – 0.30
PTFE	0.04 – 0.13
Polyimide	0.10 – 0.30

Beyond generating heat, the relative motion at the interface can produce particulates, especially during Phase I. This is caused by the fact that all surfaces, even apparently smooth surfaces, have asperity peaks on the microscopic level, see Fig. 9.4. These peaks tend to fracture from the surface because of the shearing motion. The fracture peaks can then be expelled from the interface resulting in particulates. Hard plastics, such as PC, tend to produce more particulates because they tend to fracture instead of deforming locally. In contrast, softer materials such as PP and PE tend to deform locally and produce less particulates under similar conditions. That does not mean that PP and PE do not produce particulates, only that they have a lower tendency to produce particulates with friction welding compared to harder plastics.

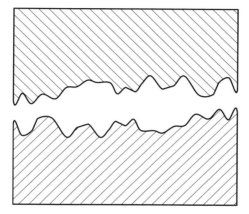

Figure 9.4 *Hypothetical micrograph of asperity peaks at weld interfaces*

During the subsequent shear melting stage (Phase II), the energy put into the weld area dominates the melting process, with the weld interface heating faster than the heat can be conducted away into the solid material surrounding the joint. Thermoplastic materials are natural insulators and are therefore poor conductors of heat. Due to the poor heat conduction, the temperature at the interface rises rapidly causing a thin layer of molten thermoplastic material (typically a few hundred microns thick) to form. Once the thin layer is formed, the process moves into the third and final phase of the vibration cycle. At this stage the shear energy equals the dissipated heat flow and the temperature in the melt is almost constant.

Once Phase III starts, indicated by a relatively linear gradient between melt displacement and time, there is no solid/solid frictional heating. Instead, heat generation is dominated by shearing. In Phase III, the process can be modeled by two solid interfaces separated by a molten layer with a thickness of 2 h, see Fig. 9.5.

Based on squeeze flow, it is possible to estimate the force required to shear the two interfaces past each other using Eq. 9.5. The first parameter is the viscosity of the melt, η. This is a measure of how easy the melt flows and it is a reciprocal function to the melt flow index. In addition, viscosity is complicated by the fact that it is dependent on the temperature of the melt and its shear rate, τ. Thus, this parameter is very complicated to model, even with a FEA technique. The other parameter effecting the heat generation in Phase III is the melt thickness, 2 h. [9]

$$F = \frac{\eta v}{2h} bd \qquad (9.5)$$

η = melt viscosity
b = width of weld interface
d = depth of weld interface

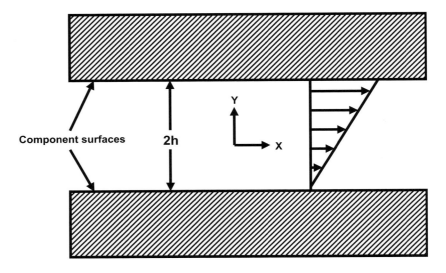

Figure 9.5 Cross section of relative velocity in vibration welded joint

With the resistance force defined, it is possible to use Eq. 9.1 to estimate the power generated during Phase III for orbital and linear welding as shown in Eq. 9.6.

It is interesting to note that there is a significant difference in the relationship between the process parameters and the amount of heat generated between Phase I and Phase III. For example, during Phase I, the orbital process dissipates 64% more power and 100% more power during Phase III compared to linear friction welding [10]. It also important to note that during Phase III, heat dissipation is dependent on melt viscosity and melt thickness. This is particularly important since both of these factors are dependent on weld line temperature. For example, the melt viscosity for most plastics is inversely proportional to temperature and accordingly, as the weld line temperature increases, the efficiency of heat dissipation decreases. This effectively makes the process self-temperature regulating, virtually not allowing for material heat degradation at the welding interface. At this stage in the welding cycle, material from the melt zone is moved to the edges of the weld where it remains as weld flash. Tests have shown that prolonging the welding cycle beyond the second heating stage and into the third phase has no additional benefit in regard of the weld strength [11]. This effectively means that in Phase III the vibration cycle can be terminated and the cooling phase commenced.

LINEAR

$$\frac{P}{Area} = (F \cdot v)$$

Over multiple cycles the average heat generated per unit area is $\dfrac{\eta A^2 \omega^2}{4h}$

ORBITAL

$$\frac{P}{Area} = (F \cdot v)$$

$$P = (\frac{\eta v}{2h} bd) \cdot v = \frac{\eta A^2 \omega^2}{2h} \qquad (9.6)$$

When the vibration ends, the cooling stage of the process begins. At the point in time when the vibration, and hence heat input, is terminated, the thermoplastic remains molten for a short while (typically less than one second). During this period, there is continued displacement of the material until it has solidified.

The four phases of the linear and orbital vibration welding process can be observed by the use of displacement transducer and a chart recorder. The profile of the curve will be dependent on the welding parameters used to produce the weld. The welding parameters will be a combination of time, pressure, amplitude, and frequency. In general, increasing the welding pressure will decrease the time taken to achieve the steady state (Phase III) and will increase the slope of the Phase III stage of the cycle. However, increasing the welding pressure too much, beyond 2 N/mm^2, will have a detrimental effect on the weld strength. Increasing the welding pressure beyond these values will result in a reduction in the melt layer thickness through shear thinning and force more material to the flash area, effectively producing a cold weld. In addition, shear thinning promotes molecular alignment parallel to the squeeze flow and reduces bridging of polymer chains across the bond line. Decreasing the amplitude will increase the time to produce the weld and decrease the slope of the steady state, Phase III heating stage.

Some work has been conducted on the use of dual stage pressure for the vibration cycle. A high pressure is used to initiate the melting process and a lower pressure is used to finish the vibration cycle. This has the overall effect of shortening the vibration cycle while increasing the weld strength.

9.4   Welding Equipment

9.4.1   Linear/Orbital Vibration Welding Equipment

The linear vibration welding process was invented in the mid 1970s as a technique for the welding of relatively large components for the automotive industry [12, 13]. A typical linear vibration-welding machine is shown in Fig. 9.6 while Fig. 9.7 shows a typical orbital machine. The linear vibration welding machine consists of a sturdy frame, sound insulation panels, a vibrating mechanism, force application system, and a microprocessor control system. The force application system will either be hydraulic or pneumatic, depending on the magnitude of force required to affect the welding process. The amount of force will depend on the size of the component being welded. The machine is built around a study frame, typically box section steel, to ensure that there is sufficient machine mass to prevent the equipment from moving during the vibration cycle. The sound insulation panels provide a protection for the operator and surrounding personnel during the welding operation, since the vibration process produces a high level of noise. Generally, the noise must be attenuated to a maximum of 80 dB for the operation of this type of equipment.

Inside the welding machine, there is a system for producing the mechanical vibrations and applying the axial force during the welding operation. The vibrations are applied via the upper tooling platen and the force is applied by the lower tooling platen. Figure 9.8 shows a schematic of the internal mechanism for producing the mechanical vibrations by electromagnets. An aluminum or steel platen is used to support the upper component tooling. Two rows of stiff, vertical springs are positioned along either side of the platen. These are fixed

Figure 9.6 Linear vibration welding machine (Courtesy Branson Ultrasonics Corp.)

to the inside of the top frame and to the platen. At the sides of the tooling platen, there is a mechanism for vibrating the platen during the vibration cycle. This will be either a hydraulic cylinder or an electromagnetic driver (as shown in Fig. 9.8), although for the majority of commercially available equipment, the electromagnetic driver system is employed. In both cases, the drive mechanism responds to an electrical signal in the form of a sinusoidal wave. The frequency of the signal can either be fixed or variable and is typically in the range 50 to 120 Hz. In theory, other input waveforms, for example a square wave, could be used although there may be little benefit to the welding operation. At the completion of a weld cycle, the driving mechanism is switched off and the springs return the tooling platen to the central position. Due to the spring stiffness, good part alignment is achieved with linear and orbital vibration welding. The part alignment is typically ±0.1 mm on either side of the pre-weld aligned position.

The motion in orbital welding is generated as seen in Fig. 9.9. The vibrations are applied via the upper tooling platen. This tooling plate is mounted on a multitude of vertical, stiff springs, each positioned at 120° relative to each other. Between each of the vertical springs, three electromagnets are positioned, again at 120° spacing relative to each other.

The vibration/orbital welding head is a resonating system and as such will only provide maximum efficiency, amplitude, and operating life if used at resonance. A welding machine may have a range of frequencies within which it is capable of operating. The mass of the tooling attached to the vibrating platen will dictate the specific frequency at which the machine will operate. On some equipment, the frequency is tuned manually using a frequency varying potentiometer and a digital visual display of the vibration amplitude. In order to achieve resonance, the frequency is manually adjusted until the current or power dissipation is minimized at a particular amplitude. Again, with the introduction of micro-

Figure 9.7 Orbital vibration welding machine (Courtesy Branson Ultrasonics Corp.)

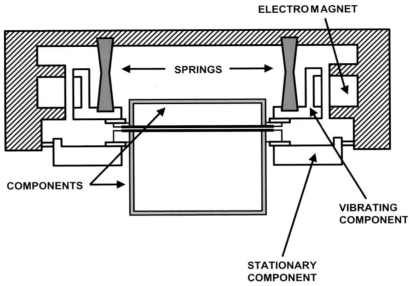

Figure 9.8 *Schematic of linear vibration mechanism*

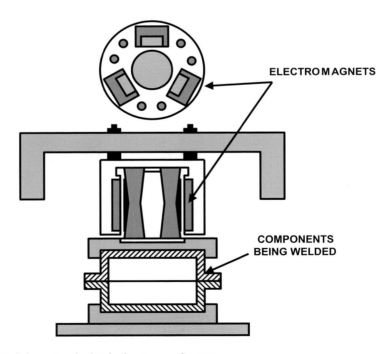

Figure 9.9 *Schematic of orbital vibration mechanism*

processor control into welding systems, automatic tuning of the welding system is now possible and is available on the majority of commercial welding equipment. To achieve tuning, the operator attaches the tooling to the upper platen and initiates the automatic tuning sequence. The machine scans a range of frequencies until the current or power dissipation is minimized and fixes the machine operating frequency to the desired level.

The base tooling platen is mounted on machined guide rails in the lower part of the machine. The number of these guide rails can vary from four (one at each corner), two (one at opposite corners), or more commonly two with one on each side, allowing the tooling plate to move in the vertical axis while applying the welding force and maintaining horizontal positional accuracy. As with the upper tooling platen, the lower platen can be either aluminum or steel. In both the upper and lower tooling platens, the component tooling will be fixed to the platen surface using steel bolts and additional dowel pins for location accuracy. Dowel pins can also aid in positioning the upper and lower tool relative to each other. It is important when designing tooling for the vibration welding process to minimize moving parts, such as pneumatic clamps, in the upper tooling platen, because these devices will experience excessive wear and can fail. It is common for pressure bladders, such as hoses, to be designed into the upper tool to help with pressure application and part fixturing (location retention). In tool design, it is important to consider that the upper/vibrating tool may experience more than 90 g's of acceleration (882 m/s^2) at 100 Hz and 200 g's (2352 m/s^2) of acceleration at 240 Hz. The vibration/orbital-welding machine may therefore contain a built-in vacuum pump to provide a vacuum to hold components in the upper tooling prior to starting the welding cycle. Figure 9.10 shows a photograph of an upper vibration welding tool used to assemble intake manifolds which has moving parts.

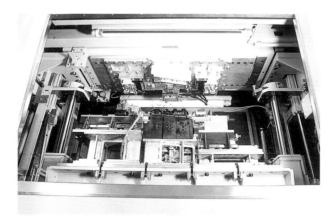

Figure 9.10 *Upper vibration tool with moving parts (Courtesy Branson Ultrasonics Corp.)*

The microprocessor control system will monitor and control the welding parameters as well as manage hydraulic pressures, cooling systems, and vacuum pumps if applicable to the equipment. Microprocessor systems may also have the capability to interface with personnel computers to down-load welding parameter information and welding cycle data for SPC (Statistical Process Control) software.

9.4.2   Component Tooling

Correct tooling design is an important aspect of both the linear and orbital vibration welding processes. Components that are welded by these processes must be adequately supported to prevent movement in the component during vibration. In the case of the linear vibration welding process, there should be no movement parallel to the vibration direction. For the orbital vibration process, it is important that there is no movement in both directions of the component. Prevention of component movement is important because the relative amplitude between the component welding surfaces will be reduced if the components are allowed to move. In addition, movement between the fixture and parts can produce marking on the part surface. For example, if a vibration amplitude of 1.8 mm (peak to peak) is required to weld a component, then a movement in the tooling of 0.3 mm will effectively produce only a relative amplitude of 1.2 mm, which may be insufficient to produce the heating and melting of the component welding-interface.

The design of the tooling nest and material selection should take into account the component complexity and the material. Injection molded components generally shrink following the injection molding process. It is, therefore, important that the design takes into consideration molding shrinkage. This can be achieved in one of two ways, either by using a split block tool in which individual segments can be precisely positioned once the final production molding is produced, or by machining a solid tool to the final production component geometry. The tools can be either steel or aluminum, with the latter being preferred due to a general requirement to save tooling weight, particularly for the upper component tooling. In some applications, an epoxy molding resin can be used to produce a precisely fitting tool around the component. These types of tool can be useful when the component shape is very complex. However, it should be noted that epoxy resin tools are prone to wear leading to movement of the component in the tooling and loss of relative movement as described previously.

To support the component correctly in linear and orbital vibration welding, the upper and lower tooling should intimately contact the component to provide good support, particularly directly adjacent to the weld areas. This ensures even pressure being applied to the welding interface during the welding process. The area of the tool adjacent the weld is often knurled or serrated to grip the component. Flexible component materials or thin walled components that require further support should have U-flanges that locate in the tooling (see Fig. 9.11). The U-flange engages the component to the tool during the vibration process, thus reducing the possibility of component movement and loss of relative amplitude. Where possible, clamping should be added to the lower component to further prevent movement. Vacuum grooves and suction pads can be added to the upper, vibrating tooling to hold the component in position before welding. Vacuum assisted tooling is useful when the component is not pre-assembled in the lower tooling before welding.

In linear and orbital vibration welding, the mass of the upper tooling is critical to achieving the correct amplitude and operating efficiency. The vibrating mechanism is a resonant system. The frequency at which maximum amplitude occurs is governed by the tooling mass. As the mass is changed, the frequency at which maximum amplitude occurs will change. To ensure long-term performance of the vibrating system, the upper tooling mass should be such that the machine operates within the manufacturers recommended frequency band. This is achieved by removing metal from non-critical areas of the tooling, i.e. not from areas adjacent to weld areas, or by adding additional mass. Generally, manufacturers provide charts relating tool mass and frequency for the maximum amplitude.

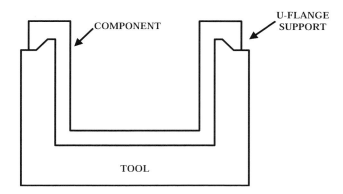

Figure 9.11 *U-Flange support – vibration welding*

9.5   Joint and Component Design

Joint and component design are critical factors when designing injection molded components for linear and orbital vibration welding [14, 15, 16]. Care should be taken to ensure that the joint achieves the desired result, for example, produces an aesthetically pleasing appearance, or provides appropriate weld strength. Component design is similar for both processes with some variation in the joint design for orbital welding compared to linear vibration welding. As with tooling design, component design is critical to ensure that the component does not move relative to the fixture during the welding process.

9.5.1   Joint Design – Designs without Flash Containment

Joint design is concerned with the part of the component immediately surrounding the welding interface. Good joint design will provide the required strength, aesthetic appearance, and component support during welding. In both the orbital and linear vibration welding techniques, circular and non-circular components can be accommodated. Joint designs can range from simple butt joints to the U-flange joints with a flash trap, each fulfilling a particular role depending on the complexity of the component and the thickness of the component wall.

The simplest joint design for both vibration and orbital welding is the butt joint. Figure 9.12 shows a butt joint between straight injection-molded plaques. In the diagram, two vibration directions are shown. In the example on the left, the vibrations are parallel to the direction of the molded plaque and in the example on the right the vibrations are transverse to the molded plaque. To ensure that the component walls remain in contact during welding, the amplitude of the transverse vibration (peak to peak) must be less than 90% of the component wall thickness. The majority of linear vibration welded applications will be enclosures. In the simplest form, they can be a square or a rectangular box component. Around the wall of an enclosed component, a combination of the transverse and parallel motion, shown in Fig. 9.13, will be present. Side A is where the motion is parallel and side B where it is transverse. With this combination of transverse and parallel motion relative to the component, wall flexing in the transverse direction can prevent welding around the entire component. It can also lead to distortion, particularly when some areas of the component melt and others do not. Note that the butt joint provides no method for securing the component during the vibration cycle, and hence is more useful for the orbital vibration process where the wall flexing is less of an issue.

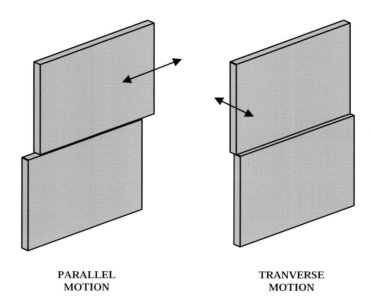

PARALLEL   TRANVERSE
MOTION   MOTION

Figure 9.12 Simple butt weld for linear and orbital vibration welding of a molded plaque

A cross-section of the simple butt joint is illustrated in Fig. 9.14, before and after welding. The illustration indicates the position of the weld flash generated during the welding process. This joint design is used where weld flash is acceptable in the appearance of the final component, for example, in an application where the product is hidden within a further enclosure when finally assembled.

In the majority of linear and orbital vibration applications, the component wall is thickened in the area of the weld. This serves two purposes: First, it gives the part greater stiffness, therefore, reducing the wall flexing during the welding process. Second, the overall component strength will be increased. In many materials, the weld strength will generally

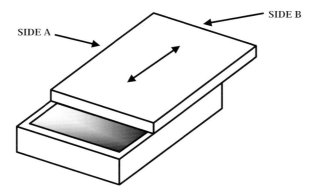

Figure 9.13 *Rectangular component showing motion around the weld interface*

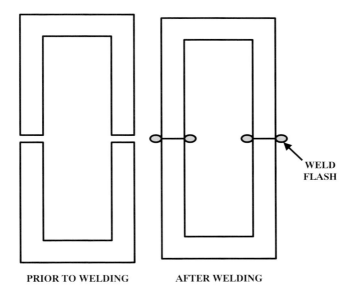

PRIOR TO WELDING   AFTER WELDING

Figure 9.14 *Simple butt joint for linear and arbital vibration welding*

be less than the parent material strength. Ideally, the force required to break the weld should be equal to the force required to break the component parent material. Therefore, by increasing the weld area, the force required to break the weld can be increased to match the force required to break the parent material. If the force required to break the weld exceeds the force required to break the parent material, failure will occur in the component and not in the weld. Two examples of the thicker wall butt joint are shown in Fig. 9.15. Either joint design could be selected depending upon the required appearance of the component. In practice, the wall thickness is generally increased between two and three times the thickness of the overall component wall, depending on the strength required from the joint.

Figure 9.16 illustrates the thicker wall butt joint and shows the dimensions required for the linear and orbital vibration welding processes. The dimension T is the thickness of the component wall and W is the thickness of the weld. Generally, a wall thickness of the component should be between 2 and 4 mm. A radius, R, is added at the internal corner to reduce the possibility of cracks initiating from this area while the component is in service. This radius dimension should be approximately 10% of the component wall thickness.

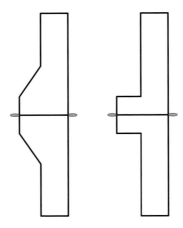

Figure 9.15 *Simple butt joints with thicker in weld area*

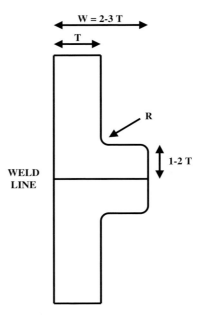

Figure 9.16 *Increased wall thickness of component to give increased component strength*

To give increased rigidity to a welded component and to secure the component into the welding tool, the U-flange joint (see Fig. 9.17) is used. It is suitable for thin-walled components or components with long, unsupported walls. The U-flange engages into a mating section of the component tool, preventing movement of the wall during the vibration process. This is particularly important when the movement is transverse to the component wall. The U-flange should be extended around the full periphery of the component. The dimensions of the flange are shown in Fig. 9.17, where T is the component wall thickness and the radius, R, should be approximately 10% of the wall thickness.

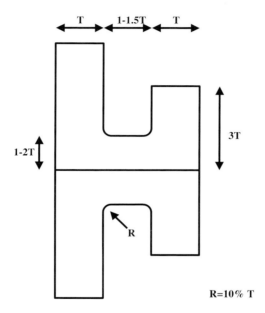

Figure 9.17 U-Flange linear and orbital welding joint design

9.5.2   Joint Design – Designs with Flash Containment

To avoid welding flash in visible areas and to produce an aesthetic joint, a flash trap needs to be added to the joint design. Two examples are shown in Fig. 9.18. The design shown in Fig. 9.18(a) is a functional design. This facilitates simple injection mold design and leaves the flash in the small groove after welding. Although this is a functional flash trap design, the weld remains visible after the welding process. Some materials, for example nylon 6, will naturally 'roll up' during the vibration cycle and will remain within the bounds of the functional flash trap. Some materials will produce flash that extrudes from the weld in a straight, flat fashion. This type of flash will not easily be contained within the functional flash trap and, therefore, an alternative flash trap design should be used. The design shown in Fig. 9.18(b) ensures an aesthetic joint appearance after welding. When the material is displaced during welding, it is enclosed in the round flash trap. Unlike the functional flash trap design, the weld is hidden after the gap has closed. The magnitude of the gap (denoted as 'a') prior to welding must be sufficient to close the gap once the weld cycle is complete. For example, if 1 mm of material is to be displaced during welding, the gap should be 1 mm. In both designs shown, the flash trap can be included either on the inside or on the outside of the component. The choice between the two types of flash trap shown will depend upon the material behavior during welding.

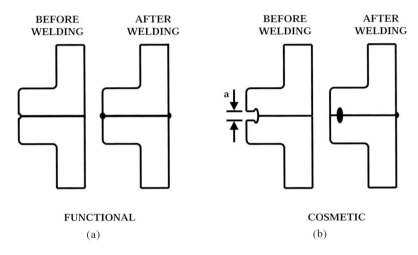

Figure 9.18 Weld flash trap designs

A commonly used, alternative joint design is the tongue and groove design shown in Fig. 9.19. This design traps the flash within the groove inside the wall rather than at the outside edge of the component as in the two previous designs. This will ensure that the flash will be contained independently of material flow during the vibration cycle, i.e., rolled or flat squeeze flow. The addition of radii, R, is recommended as shown previously in Figs. 9.16 and 9.17. These radii should be 10% of the wall thickness of the component. The width of the tongue should range from one to three times the wall thickness of the component when the vibration direction is parallel to the component wall. The actual width will depend on the strength requirements of the joint. A gap, not greater than half the component wall thickness, should be incorporated in the joint design on either side of the tongue. This gap will accommodate the weld flash as it is generated. When the vibration is transverse to the component wall, there should be an additional allowance in the groove width to accommodate the vibration movement during welding. A gap (denoted by 'a') must allow for the amount of material displaced during welding, otherwise welding will occur around the edge of the flash trap.

Some thermoplastic materials tend to move upwards and out of the flash trap before the flash trap closes. To compensate for this, the tongue and groove joint design can be modified by incorporating an additional tongue in the lower groove of the joint, which is around 75% of the width of the upper tongue. This will automatically direct the welding flash downwards away from the gap, allowing the flash trap to close before the flash leaves the joint area. This modified joint is shown in Fig. 9.20.

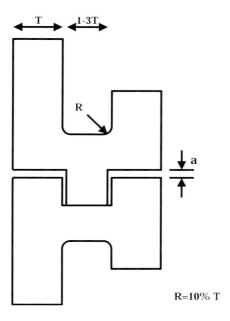

Figure 9.19 *Tongue and groove linear and orbital vibration welding joint design*

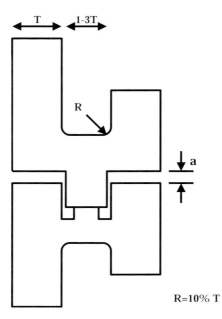

Figure 9.20 *Tongue and groove linear and orbital vibration welding joint design. Additional lower component tongue directs flash downward*

Figure 9.21 shows a variation in the groove profile to improve part alignment by locking the mating component and tool together. This design is suitable for components that show a variation in the shrinkage from component to component. Figure 9.22 shows a joint design with rib stiffeners that can help prevent component deflection during welding.

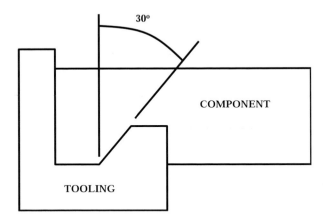

Figure 9.21 *Joint design for locking the component into the machine tooling*

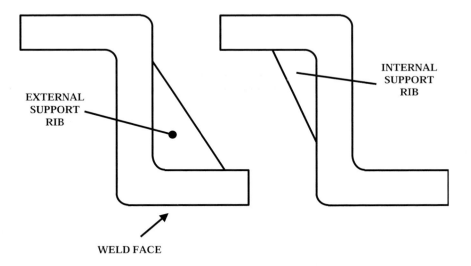

Figure 9.22 *Additional component strength gained using a) internal (shown on left) or b) external ribs (right).*

9.5.3   Selecting the Most Suitable Joint Design

The selection of the weld joint design will depend on the desired component performance, i.e. strength, its appearance, and its functionality. Table 9.4 provides a selection guide based on the joint requirements. It is important to note that the final joint strength will be determined by the material weldability, application design, and welding process parameters.

Table 9.4 Linear and Orbital Vibration Welding Joint Selection Guide

		Appearance ← →		
		Not Critical		Critical
Strength ↑ Not Critical		Simple butt joint, see Fig. 9.14.	Flash traps not easy to incorporate into single butt joint without thickening wall.	
Critical ↓		Simple butt joint with thicker wall in weld area, see Fig. 9.15	Simple butt joint with thicker wall and functional flash trap, see Fig. 9.18.	Simple butt joint with thicker wall and aesthetic flash trap, see Fig. 9.18.
			Internal wall flash trap with U-flanges for extra component support during welding, see Fig. 9.19.	

9.5.4   Component Design

An important aspect of vibration/orbital welding is the design of the component, which should not be confined to the functionality of the application but should facilitate the welding operation. For example, when a component has a thin supporting wall and is liable to movement during the vibration cycle, ribs should be added to provide rigidity (see Fig. 9.22). The ribs can be added adjacent to the joint area as in example (a). This reduces the effects of offset loading caused by a wide tongue and groove joint (shown previously in Fig. 9.19). In example (b), the ribs are added on the inside of the component to give the overall part rigidity. Generally, the thickness of the ribs would be between 50 and 100% of the component wall thickness. Typically, the ribs would be 45° to the wall of the component and, therefore, governed by the width of the flange available at the base of the rib.

For linear vibration welded components, the component wall must be able to support the component under welding compression loads (0.5 to 2 N/mm^2) and support sideways vibration during welding. U-flanges give further support and rigidity to the perimeter of components as well as providing tooling location. When thin walls present flexing problems during welding, particularly when vibrating across the component wall, stiffening ribs should be added either internally or externally, adjacent to the weld flange area.

The joint perimeter should be flat, although because of the high clamping forces used for this process, some warpage can be tolerated. To assure high quality welds, it is ideal to apply the weld pressure directly behind the joint. This should be taken into account when designing components and positioning the joint. Finally, molded pins and holes on the mating component can be used to locate the components together prior to welding. During the weld cycle, these pins will break away to allow for the relative motion between the two parts. Pre-assembly can be carried out when the operator is waiting for the current weld cycle to be completed. Typically, the pins would be 1mm in diameter.

9.6   Applications

Applications of the linear and orbital vibration welding processes vary greatly, although they are predominately automotive. Applications are restricted to single plain motion and must be able to tolerate relative motion. Applications include polycarbonate air diffuser for aircraft, acetal boiling-chamber for water [13], car bumpers, air intake manifolds, dashboards, and glove box compartment components.

One of the first applications assembled with vibration welding were carbon canisters [17]. Another early application of vibration welding was the car bumper [12]. Two thermoplastic sections, a skin and a stiffening bar, were welded using linear vibration welding in around 8 to 10 s, providing for significantly increased production output. Other large-scale components, such as instrument panels and dashboards assemblies are now routinely manufactured using the linear vibration welding process.

A further, major application of the linear vibration welding process is the automotive air intake manifold [18]. The potential of thermoplastics in this application was first introduced in the mid 1980s. Injection molded nylon air intake manifolds offer greatly improved engine performance due to smoother air flow and reduced weight. Original designs were manufactured using the fusible core molding process (lost core technology). Although this process is useful for very complex manifold designs, the capital cost of the equipment is prohibitive for the majority of applications. Linear vibration welding in this application revolutionized the manufacture of the intake manifold by reducing the overall capital cost significantly. Two or three injection molding machines are employed to produce the component parts of the manifold and linear vibration welding is used to assemble the final manifold. Typically, welding cycle times range from 25 to 30 s. Figure 9.23 shows a vibration welded air intake manifold and Fig. 9.24 shows a vibration welded automotive dashboard.

Figure 9.23 *Vibration welded automotive air intake manifold (Courtesy of TWI)*

Figure 9.24 *Vibration welded automotive dashboard (Courtesy of Branson Ultrasonic)*

Applications of the orbital vibration welding process include rear light clusters for cars, where an ABS housing is welded to a polycarbonate lens. As with linear vibration welding applications, the vibration time is short, typically 4 to 6 s, with a complete machine cycle time of 20 to 30 s. Other applications include domestic products, such as irons.

9.7   **Material Weldability**

One of the major advantages of the friction welding processes over other thermoplastic welding processes is the ability to weld almost all thermoplastics. The exceptions are fluoropolymers because they tend to have relatively high molecular weight and low coefficient of friction compared to the majority of thermoplastics. In general, materials with high melting points will require longer welding times to achieve heating. In some materials, it can be difficult to weld combinations of like materials where one component part is manufactured in an extrusion grade and the other in an injection molded grade. This is due to differences in melt flow and melt temperature. Table 9.5 provides the weldability for most common thermoplastics.

Table 9.5 Weldability of Common Thermoplastics (Friction Processes)

Material	Weldability
ABS [19]	Excellent
Acetal	Excellent
Acrylic	Excellent
ASA	Good
Cellulose-Acetate	Excellent
CTFE	Good-Fair
ECTFE	Good-Fair
EVA	Good
FEP	Poor
Polyethylene	Excellent
High impact polystyrene	Excellent-Good
PBT [20]	Excellent
PEEK	Good
PES	Good

Material	Weldability
PET	Excellent
Polyamides (nylon) [5, 8, 18]	Excellent-Good
Polycarbonate [21]	Excellent
Polypropylene [17, 21]	Excellent
Polystyrene	Excellent
Polysulphone	Excellent
PPO [19]	Excellent
PPS	Good
PTFE	Poor
PVC	Good
PVDF	Good
PVF	Good
SAN	Excellent

There are a few combinations of dissimilar thermoplastic materials that are compatible for welding, although there are no combinations where an amorphous material is compatible with a semi-crystalline material. The thermoplastics that are compatible include ABS to polycarbonate, ABS to acrylic, SAN to acrylic, SAN to polystyrene, and polycarbonate to acrylic. Table 9.6 lists achievable weld strengths between selected plastics. It is important to note that these are possible achievable strength and individual results may vary depending on application and equipment set-up. The table lists the tensile strength of base material and percent strain at failure. The achievable weld strength is given in terms of weld factor (weld strength/base material strength (see Chapter 15)) of the weaker material when dissimilar materials are joined. The number below the weld factor is the percent strain at failure [19].

Table 9.6 *Compatibility and Possible Weld Strength for Selected Plastics with Linear Vibration Welding (Courtesy V. K. Stokes GE Plastics)*

	ABS 44 MPa (2.2%)	ASA 32.5 MPa (2.9%)	M-PPO 45.5 MPa (2.5%)	M-PPO/ PA 59 MPa (>18%)	PC 68 MPa (6%)	PBT 65 MPa (4.5%)	PC/ABS 60 MPa (4.5%)	PC/PBT 50 MPa (-)	PEI 119 MPa (6%)
ABS 44 MPa (2.2%)	0.9 (2.1%)		0.76 (1.45%)		0.83 (1.7%)	0.8 (1.6%)	0.85 (1.8%)		0.65 (1.14%)
ASA 32.5 MPa (2.9%)		0.46 (0.9%)							
M-PPO 45.5 MPa (2.5%)	0.76 (1.45%)		1.0 (2.4%)	0.22 (0.35%)	0.24 (0.4%)				0
M-PPO/ PA 59 MPa (>18%)			0.22 (0.35%)	1.0 (>10%)	0.29 (0.75%)				
PC 68 MPa (6%)	0.83 (1.7%)		0.24 (0.4%)	0.29 (0.75%)	1.0 (6%)	1.0 (1.7%)	0.7 (1.8%)	1.0 (4.9%)	0.95 (2.75%)
PBT 65 MPa (4.5%)	0.8 (1.6%)				1.0 (1.7%)	0.98 (3.5%)			0.95 (4.1%)
PC/ABS 60 MPa (4.5%)	0.85 (1.8%)				0.7 (1.8%)		0.85 (2.3%)		
PC/PBT 50 MPa (-)					1.0 (4.9%)			1.0 (>15%)	
PEI 119 MPa (6%)	0.65 (1.14%)		0		0.95 (2.75%)	0.95 (4.1%)			1.0 (6%)*

*High strengths can only be achieved with 250 to 400 Hz welding machines/ (1 MPa = ! N/mm^2)

9.7.1   Fillers

Fillers, such as glass or talc, are added to a thermoplastic material to alter the properties of the parent material. For example, glass fibers may be added to alter the overall strength of the material. The weldability of these materials will be affected by the addition of the filler material. With a high percentage (30%) of glass filler, the amount of polymer available for welding at the interface is reduced. Consequently, a reduction in the weld strength compared to the parent material can occur. When welding raw material (no filler), for example polyethylene to polyethylene, joint strengths close or equal to the parent material can be achieved. Once a filler is added to the parent material, the overall strength of the material increases, but the strength of the weld will remain similar to the base polymer. The maximum theoretical weld strength that could be achieved would be that of the parent base polymer. For example, linear vibration welds in 30% glass filled nylon 66 achieve maximum weld strengths of 70% of the parent material [11]. To compensate for this reduction in weld strength, designers increase the thickness of the weld area using one of the joints described in Section 9.5. Some studies have indicated that it may be possible to get a small portion of the glass fibers to bridge across the weld line during the welding process [20, 21]. This would allow the weld strength to exceed the base material strength of an unfilled polymer. This effect has only been seen in materials with relatively low amounts of glass filler, up to 10%, and only when enough glass crosses the weld line to increase the weld strength approximately 10% above the unfilled base material.

In addition to affecting welding quality, glass fillers and hard materials, for example PPS, can wear the component support tooling. The condition of the component support tooling should be checked regularly for signs of wear.

Additives in materials such as silicones and other lubricants will reduce the friction coefficient of the material, which will make them more difficult to weld. This can be particularly relevant when materials are based on fluoropolymers, for example FEP where the coefficient of friction is relatively low in the base polymer. Other additives, for example plastizers may also affect the properties of the weld. Additives can affect the coefficient of friction, shear heating mechanisms and migrate and collect at the weld interface.

Moisture absorption by the polymer will also affect a material's weldability. A reduction in weld strength up to 20% has been observed in saturated (50% relative humidity) nylons compared to 'dry as molded' nylons. It is recommended to weld materials that are hygroscopic as soon after molding as possible. If this is not practical, for example when the component parts are manufactured in different locations, then the components should stored in a dry environment or dried in an oven following the material manufacturer's guidelines. The most prominent effect of moisture on weld quality is the formation of porosity. As the weld line temperature increases during the welding process, the moisture tends to boil resulting in a corresponding vapor pressure. If the vapor pressure exceeds the weld line pressure (see Section 9.8.6), the moisture can form porosity depending on the amount of time, moisture content, and material [22].

9.8   Equipment Setup

9.8.1   Welding Parameters and Welding Parameter Setup

For all thermoplastic welding processes there are three main parameters: heat, time, and pressure. The way in which each parameter is controlled depends on the welding process. Individual equipment manufacturers may offer variations or describe the parameters in different terms. For example, welding a component by displacement may be described as welding by distance or by way. In linear and orbital vibration welding, the heat is generated by the friction at the weld interface. Welding pressure is generated by the fixtures and clamping mechanisms and is conveyed through the component to the weld interface. Time is controlled by a microprocessor controller. It can not only define the actual heating (vibration) time but can be limited to a pre-selected the weld material displacement. It is recommended that Design of Experiment (DOE) techniques are employed when determining the welding parameters.

9.8.2   Vibration Amplitude and Frequency

The vibration amplitude is the horizontal peak-to-peak movement of the vibrating welding head. The frequency is the number of vibration cycles per second. The frequency and the amplitude are related. In terms of typical machine design, doubling of the vibration frequency will result in a halving of the vibration amplitude and vice versa. Typically, a vibration frequency of 200 Hz has a vibration amplitude of 1 to 1.8 mm while a frequency of 100 Hz will have an amplitude of 2 to 4 mm. The lower frequency/higher amplitude tends to be used for components with flexible supporting walls and large joint areas, for example, car body components such as bumpers. The lower amplitude/higher frequency tends to be used for smaller applications, for example, automotive air intake manifolds.

The method for setting the vibration amplitude will vary from machine to machine. It can be achieved either by varying the tooling mass on the vibrating head or electronically. In either case, it is important that the vibrating system operates at resonance to ensure optimum machine performance and longevity. Manufacturer guidelines should be observed when setting up new tooling.

Orbital vibration welding operates at lower amplitudes than linear vibration welding. The amplitude is typically 0.25 to 1.5 mm. Setting the vibration amplitude is similar to setting the amplitude in the linear vibration welding process.

9.8.3   Weld Force/Pressure

The welding force and welding pressure, although related, are separate entities. The welding force is the amount of force applied to the component during welding. The welding pressure is the amount of pressure applied at the joint during welding. Equation 9.7 explains the relationship between welding force and pressure.

$$\text{Welding pressure (N/mm}^2 \text{ or MPa)} = \frac{\text{Welding force (N)}}{\text{Weld area (mm}^2)} \qquad (9.7)$$

Typically, welding pressures range from 0.5 to 2.0 N/mm^2. Increasing the weld pressure beyond these values can reduce the strength of the weld by forcing out all the molten plastic, resulting in a 'cold' weld (see Section 9.3). For example, in Nylon 6 and 66 the weld strength is reduced by up to 40% when increasing the weld pressure from 1.0 to 20.0 N/mm^2.

Dual stage welding pressure can have a positive effect on weld quality and shorten the overall weld time. When welding with dual stage pressure, a higher welding pressure is used to start the interface material melting and a lower pressure is used to complete the vibration cycle, resulting in the optimum weld strength. Table 9.7 provides approximated values for the welding pressure of different materials.

Table 9.7 Welding Pressure for Thermoplastic Materials (Approx. Values)

Material	MVR [cm³/10 min]	Welding pressure $\frac{N}{mm^2}$
ABS	2–50 (220/10)	1.0–2.0
PA6	18–110 (275/5)	0.5–3.0
PA66	10–180 (275/5)	0.5–3.0
PC	3–12 (200/1.2)	1.0–2.0
HDPE	0.1–80 (190/5)	0.5–8.0
PMMA	1–25 (230/3.8)	1.0–2.0
POM	1–40 (190/2.16)	1.0–4.0
PP-H	1–100 (230/2.16)	0.5–4.0
PPE+SB	8–270 (250/21.6)	2.0–6.0
PS	3–25 (200/5)	1.0–4.0
SAN	8–25 (220/10)	1.0–2.0

MVR = melt viscosity Range

9.8.4   Cool Time

The cool time, or sometimes known as the hold time, is the length of time at which the parts remain under pressure after the vibration cycle is completed. It should be sufficient to allow the material to cool and solidify but no longer as this would extend the overall weld cycle time. It is generally determined by experimentation, but is typically in the region of 4 to 10 s. When initially determining the vibration time, the cool time should be set longer than necessary in order to assure that the weld strength results are not influenced by the cool time.

9.8.5   Welding Modes

9.8.5.1   Welding by Time

In linear and orbital vibration welding, the components can be welded using a set weld time. The weld time is the amount of time that the plastic parts are rubbed together to create the heat necessary for welding. The weld time should be sufficient to allow molten material to be produced at the joint interface. In general the higher the melting point of the material, the longer the required weld time. The weld time should ideally be terminated when the steady state phase of the weld cycle is achieved. This occurs when the rate of material displacement against time becomes linear (see Section 9.3). In practice, this can be monitored using a displacement transducer and a chart recorder. Typically, the vibration welding process will take around 2 to 6 s to achieve the steady state. However, for some complex component joint designs reaching the steady state phase may not always be obvious on the chart recording.

9.8.5.2   Weld by Melt Depth/Displacement

An alternative to welding by a fixed time, is welding by displacement/melt depth. The vibrations are applied to the components until a fixed depth of material displacement, for example 1mm, is achieved. The displacement value selected must be sufficient to ensure that steady state melting occurs. Once the steady state melt occurs, there is no further benefit in prolonging the welding cycle since the weld will not become stronger (see Section 9.3). If the component incorporates mating surfaces that that should not be welded, but be in contact at the end of the weld cycle, then the selected displacement value must be equal to the pre-welding distance between these surfaces. This was described in Section 9.5.3 and shown in Fig. 9.18.

When welding by displacement, an allowance should be made for joint area flatness. If the joint area is distorted rather than flat prior to welding then the displacement value selected should be added to the maximum deviation from flatness. Flatness can be measured by placing the component onto a flat surface and measuring the height of the largest the gap around the perimeter.

In general, the software controlling the weld displacement will discontinue the vibration mechanism when the selected displacement is achieved. However, the material will remain molten for a short time and therefore the displacement of material will continue. In some applications, the final displacement may be critical. Therefore, the overrun must be taken into account when selecting the welding displacement. For example if the overrun on a 1 mm weld displacement is 0.1 mm, then the selected value should be 0.9 mm rather than 1 mm.

9.8.5.3   Fault Identification

Unsatisfying or unacceptable welds can occur as a result of the welding operation or as a result of the component design. In some cases, changing the welding parameters can rectify the fault. However, welding problems caused by a poorly designed component can be difficult and expensive to solve. Table 9.8 provides a list of common welding problems while Table 9.9 provides a list of common problems caused by improper component design.

Table 9.8 Process Related Welding Problems and Their Solutions

PROBLEM	SYMPTOMS	POSSIBLE CAUSE	SOLUTION
Overwelding	• Excessive weld flash • Final component dimension too small	• Weld time too long • Too much weld displacement • Poor flash trap design	• Reduce weld time or displacement • Evaluate flash trap design
Underwelding	• Poor weld strength • Final component dimension too large	• weld time too short • Insufficient displacement • Material difficult to weld due to low friction	• Increase weld time or displacement • Consider material change
Non-uniform weld around component	• Excessive flash around weld • Poor welds • Failure when leak tested	• Warped parts/poor moldings • Uneven weld interface	• Check part dimensions • Check molding process conditions
		• Lack of parallelism between fixture and part	• Shim fixture where necessary • Ensure tooling true to base • Check part dimensions
		• Wall flexure during welding	• Design parts to incorporate strengthening ribs and U-flanges
		• Insufficient fixture support (urethane fixtures)	• Modify fixture to prevent outward flexure • Improve support in critical areas

Table 9.8 *Process Related Welding Problems and Their Solutions (Continuation)*

PROBLEM	SYMPTOMS	POSSIBLE CAUSE	SOLUTION
			• Redesign fixture to improve rigidity • If large sections of urethane are deflecting, add rigid back-up • Check for part shifting during welding • Check provisions for alignment in mating parts
		• Part tolerances	• Improve part tolerances
		• Poor part alignment in fixture	• Re-dimension parts
		• Mold release agent at weld	• Check molding process conditions • Clean mating surfaces with suitable degreasing agent • Use a paint-able/printable mold release if required
Inconsistent weld results from part to part	• Part failure in service • Poor weld strength	• Fillers • Mold release agent at weld interface	• Reduce amount of filler • Check molding process conditions • Clean mating surfaces with suitable degreasing agent • Use a paintable/printable mold release if required

PROBLEM	SYMPTOMS	POSSIBLE CAUSE	SOLUTION
		• Part tolerance	• Improve part tolerances • Check part dimensions • Check molding process conditions
		• Molding cavity-to-cavity variations on multi-cavity moulds	• Run statistical study on molding process to see if pattern develops with certain cavity combinations • Check part tolerances and dimensions • Check for cavity wear in mold • Check molding process conditions
		• Regrind/degraded plastic	• Reduce percentage of regrind • Improve quality of regrind
		• Poor distribution of filler content	• Check processing conditions
		• Incompatible materials or resin grades	• Check with resin supplier
		• Moisture in molded part (usually nylon parts)	• Weld directly after molding • Dry parts before welding
		• Drop in air-line pressure • Clean mating surfaces with suitable degreasing agent	• Raise compressor output pressure • Add surge tank with a check valve

Table 9.8 *Process Related Welding Problems and Their Solutions (Continuation)*

PROBLEM	SYMPTOMS	POSSIBLE CAUSE	SOLUTION
		• Use a paintable/ printable mold release if required	
Flash	• Excessive material around weld area	• Inadequate flash trap design	• Review flash trap design
		• Overwelding	• See overwelding
		• Non-uniform joint	• Check molding process conditions
Misalignment of welded assembly	• Alignment at edges of welded parts is poor	• Wall flexure during welding	• Add ribs to molded part • Use U-flange joint design
		• Part tolerance/poor molding.	• Tighten up part and tooling tolerances • Check molding process conditions
		• Incorrectly aligned upper and lower tooling	• Realign tooling
		• Very high weld pressure	• Reduce to recommended levels • Add ribs to molded part

Table 9.9 Component Design Related Welding Problems and Their Solutions

PROBLEM	POSSIBLE CAUSE	SOLUTION
Internal components damaged	• Overwelding	• Reduce weld time or displacement • Evaluate flash trap design
	• Internal components improperly mounted, i.e., too close to joint area	• Make sure internal components are properly mounted • Isolate internal components from housing
Melting/fracture of part sections outside of joint	• Overwelding	• Reduce weld time or displacement • Evaluate flash trap design
	• Internal stresses	• Check molding process conditions • Check part design
	• Weld force too high	• Lower pressure and re-evaluate component design
Internal parts welding	• Internal parts same material as housing	• Change material of internal parts • Lubricate internal components • Consider component redesign
Marking	• Incorrect fit of part to fixture	• Check for proper support • Redesign fixture • Check for cavity–to-cavity variations
	• Movement during welding	• Redesign component to fit tooling correctly

9.9   **References**

1. Astrop, A., *Machinery and Production Engineering* (1979) **134**, p. 41–43

2. Mock, J.A., *Plastics Engineering* (1983) **39**, p. 27–29

3. Klos, W., *Kunststoffe* **84** (1994) 10, p. 1464–1469

4. Mengason, J., *Plastics Engineering* (1980) **36** No.8, p. 20–23

5. Uebbing, M., *Welding in the World* (1995) **35** No.4, p. 223–231

6. Potente, H., Michel, P., Ruthmann, B., *Kunststoffe* **77** (1987) 7, p. 711–716

7. Stokes, V.K., *SPE ANTEC Tech. Papers* (1988), p. 871–874

8. Sear, F.W., Zemansky, M.W., Young, H.D., *University Physics*, 6th Edition (1982) p. 25–33, Addison-Wesley Publishing Company, Menlo Park, CA

9. Stokes, V.K., *Analysis of Friction (spin) – Welding Process for Thermoplastics*, Journal of Materials Science, Vol. **23** (1988)p. 2772–2785, Chapman and Hall Ltd.

10. Grewell, D., Benatar, A., *A Process Comparison of Orbital and Linear Vibration Welding of Thermoplastics*, ANTEC-99, Conference Proceedings, Society of Plastics Engineers, Brookfield, CT

11. Froment, I., *Vibration welding Nylon 6 and Nylon 66 – a Comparative Study*, ANTEC-95, Conference Proceedings, Society of Plastics Engineers, Brookfield, CT

12. Robinson, I., *Welding and Metal Fabrication* (1989), p. 152–154

13. White, P., *British Plastics and Rubber* (1992) June, p. 33–34

14. DVS Guide 2217, Part 1, Nov.1992, *Welding in the World* (1995) **35** No. 2, p. 138–146

15. NN, *Design Engineering* (1978), p. 47–48

16. Panaswich, J., Paper 840223 presented at the International Congress & Exposition, Detroit, MI, February (1984)

17. Shoh, A., *"Friction Welding Device"*, US Patent 3,920,504, Nov. 1975

18. Mapleston, P., *Modern Plastic International* (June 1995), p. 23–24

19. Stokes, V.K., *Towards a Weld-Strength Data Base for Vibration Welding of Thermoplastics*, ANTEC-95, Conference Proceedings, Society of Plastics Engineers, Brookfield, CT

20. Kagan, V., *Joining of Nylon based plastic components – Vibration and Hot Plate welding technologies*, ANTEC-99, Conference Proceedings, Society of Plastics Engineers, Brookfield, CT

21. Grewell, D., *An Application Comparison of Orbital and Linear Vibration Welding of Thermoplastics*, ANTEC-99, Conference Proceedings, Society of Plastics Engineers, Brookfield, CT

22. Potente H., *The Effect of Moisture on Vibration Welds Made with PA*, ANTEC-92 Conference Proceedings, Society of Plastics Engineers, Brookfield, CT

10 Spin Welding

Paul Rooney and David Grewell

10.1 Introduction

Spin welding of thermoplastics is one of the internal frictional welding processes commonly used in industry to weld parts with round or cylindrical weld lines, although often the parts themselves are not round in shape. The process involves holding one part stationary and rotating the second part while the two parts are under a pre-load in order to assure proper alignment and heating.

The benefits of spin welding include:

- Its speed, with typical cycle times between 1 and 5 seconds.
- Relatively low capital equipment costs compared to other frictional welding processes, such as linear vibration welding or orbital welding.
- The fact this process can weld under water.
- With spin welding it is usually possible to achieve a hermetic seal.

Disadvantages of spin welding include:

- Limitation to applications with round or cylindrical weld lines.
- Limitation to materials with a relatively high coefficient of friction in order to assure proper initial heating of the faying surface.

10.2   Process Description

10.2.1   Spin Welding Process

Spin welding consists of five major steps:

1) *Loading parts.*   This usually involves manually placing the parts into the fixture. Often there is only one fixture, the lower fixture, which holds both parts. The upper part, usually the one that is spun, is captured or engaged by the upper fixture.

2) *Press actuation.*   This step is initiated by the operator and is usually accomplished by the closing of two palm buttons. Once the upper fixture, or head, captures the parts the head applies a force while the parts are spun relative to each other.

3) *Spinning.*   Once the parts and fixtures are fully engaged, the spinning or rotation motion is initiated. This promotes heating and melting of the faying surface (surface to be welded). The duration of the heating time is usually pre-selected by the operator. In some cases, the amount of melt or displacement can be pre-selected. A typical rate of rotation is 2000 rpm, however, commercial equipment can often provide rates of rotation between 200 and 14,000 rpm [1]

4) *Clamp phase/hold phase.*   During this phase, the parts are held together under a preset pressure to assure intimate contact between the molten surfaces.

5) *Part removal.*   Once the molten material is sufficiently solidified, the parts are removed.

Table 10.1 Typical Spin Welding Cycle

Step	Typical time [s]
Loading parts	2 to 5
Press actuation	1 to 2
Spinning	0.5 to 2
Hold phase	1 to 2
Part removal	2 to 5

10.2.2   Spin Welding Modes

10.2.2.1   Inertial Spin Welding Mode

Historically, spin welding has been used to weld circular parts where relative alignment between the two parts being joined is not critical. In older machines, the rotating motion was generated by either pneumatic motors coupled to a fly-wheel, or by electric motors. In either design, the motion would continue until the all the stored energy was dissipated. Thus, it was not possible to control the final position and relative orientation between the two parts at the end of the cycle. This mode of operation is simple and requires only a limited amount of control. Machine concepts as simple as drill presses have been used to weld components in this mode. Dedicated spin welding machines that operate in this mode are usually inertial driven machines. These machines have a small fly-wheel that is driven with a pneumatic engine. In some cases, appendages are added to the part design, which engage the fixture to allow mechanical coupling between the part and fixture. Once the faying surfaces are melted and the process is discontinued, the appendages are designed to fracture (break-away) as a result of resistance force as the material solidifies. This allows the spinning motion to be discontinued nearly instantaneously and enhances weld quality since the weld does not solidify under a spinning motion. An alternative method of stopping the spinning motion at the weld line during the solidification phase is to have a breaking mechanism with the equipment design. Additional information on these machine designs can be found in Section 10.4.

10.2.2.2   Drive Spin Welding Mode

In some applications, alignment between the two parts is critical and an additional level of control is required. Recently, machine designs have utilized servo-motor technology so that, once the faying surface is fully melted, the parts can be orientated within 1 degree relative to each other [2]. This is accomplished by programming the servo- or step-motor to stop rotating nearly instantaneously at a pre-selected angle. These machines are significantly more complex compared to pneumatic fly-wheel designs, however, some applications require this level of control.

In selected servo-driven machines, it is possible to program the head to rotate between two selected angles (orientations) without making an entire rotation. That means that the machine can be programmed to rotate back-and-forth between two angles, allowing applications that cannot be fully rotated to be welded with spin welding. Due to the acceleration necessary to generate such a motion, this process is usually limited to parts less than 25 mm in diameter.

10.3   Physics of the Process

10.3.1   Process Phases

Similar to other friction welding processes in plastics, spin welding can be divided into four basic phases [3]. It is customary to view or separate the phases by plotting a typical penetration (d_y) or melt down as a function of time, see Figure 10.1.

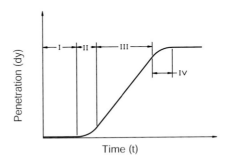

Figure 10.1 *Melt penetration as a function of time for illustration of different phases during spin welding*

While penetration graphs may vary depending on process parameters and applications, the ideal graph represented in Figure 10.1 clearly shows the four phases detailed below.

Phase I: Solid-to-Solid Friction

This is the initial phase of the weld cycle and the faying surfaces are solid. The heat generated during this phase is defined by frictional heating, see Section 10.3.2.

Phase II: Transition Phase

During this phase, the heat within the bond line is sufficient to start melting the interfaces to produce small regions or pockets of viscoelastic polymer flow. Historically, this phase has not been studied in detail because of the complexity between the interactions of the solid/solid, solid/viscous, and viscous/viscous boundaries. In addition, because this phase usually represents only a small portion of the entire welding cycle, its importance on the welding process has been considered relatively insignificant.

Phase III: Steady-State Phase

In contrast to Phase II, the steady-state phase of the welding process has been studied in great detail [4, 5], because it can be used as an indicator of weld quality and weld progress. It has been well documented that by monitoring the weld penetration and determining when Phase III has occurred, welding can be terminated without any loss in weld quality. During this phase, heat generation is limited to shearing of the viscoelastic molten layer and relatively low compared to the heat generation during Phase I. As a result, the amount of plastic material melted (softened to allow flow) within the bond line is equal to the amount of plastic material ejected out of the weld as flash. This is why this phase is often referred to as the steady-state phase; it is the phase, when melt generation equals melt ejection (see Section 10.3.3).

Phase IV: Hold Phase

This phase begins at the point when the spinning motion is discontinued. During this phase, the melt solidifies and completes the welding process. Usually, there is a small amount of additional penetration/displacement because of the clamp force that is applied to the parts to assure proper fit-up. This additional penetration is seen in Fig. 10.1.

Before the individual phases are studied in detail, the fundamental approach to modeling heat generation is based on the simple equation relating force and velocity to power, see Eq. 10.1.

$$\text{Power} = \text{Force} \times \text{Velocity} \qquad (10.1)$$

This equation shows that power is directly proportional to velocity and force. This is important to note because the velocity of a rotating body is defined in Eq. 10.2. Here ω is the angular velocity of the body in terms of Rad./s and r is distance from the center of rotation to the area of interest, in this case to the weld, see Fig. 10.2. This means that the further away a weld is from the center line (center of rotation) the higher the power generation. In other words, the larger the part, the more power generated. In addition, no heat is generated at the center line where $r = 0$.

$$\text{Velocity} = v = r\omega \qquad (10.2)$$

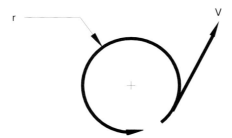

Figure 10.2 Demonstration of velocity during spin welding

It is also important to note that if the weld line is relatively thick, the heat generation on the inner diameter of the weld line will be slightly less compared to the heat generation on the outer diameter.

10.3.2   Solid-to-Solid Frictional Heating (Phase I)

As stated previously, the first phase involves shearing of two solid interfaces past each other. The absolute motion of each component is not important, only their relative motion is. As seen in Eq. 10.2, the velocity is dependent on weld location and rate of spin. As two solid interfaces are slid past each other, there is a force generated that resists the motion, called the frictional force (F_f), see Fig. 10.3.

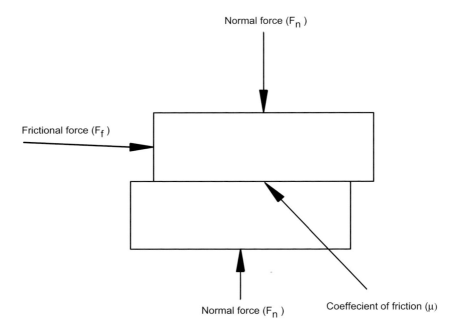

Figure 10.3 *Details of frictional forces*

The frictional force is defined in Eq. 10.3 [6], where F_n is the normal force applied to the parts and μ is the coefficient of friction between the two interfaces.

$$F_f = F_n \cdot \mu \tag{10.3}$$

Because spin welding is a process of motion, only the dynamic coefficient of friction is important to modeling the process. In general, the dynamic coefficient of friction for plastics can vary greatly. For example, an interface of PC on PC can have a dynamic coefficient of friction as high as 0.45, while the Teflon/Teflon® coefficient is only 0.04.

Some materials, such as Teflon® have such a low coefficient of friction that they are not well suited to spin welding because the amount of heat generated through frictional heating is not sufficient.

Table 10.2 *Typical Values of Coefficent of Friction for Selected Plastics [7]*

Material	Coefficient of friction (μ)
Nylon	0.20–0.40
Acetal	0.15–0.30
PTFE	0.04–0.13
Polyamide	0.10–0.30

Beyond generating heat, the relative motion at the interface can produce particulates, especially during Phase I. This is caused by the fact that all surfaces, even apparently smooth surfaces, have asperity peaks on the microscopic level, see Fig. 10.4. These peaks tend to fracture from the surface because of the shearing motion. The fracture peaks can then be expelled from the interface resulting in particulates. Hard plastics, such as PC, tend to produce more particulates because they tend to fracture instead of deforming locally.

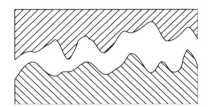

Figure 10.4 *Hypothetical micrograph of asperity peaks at two interfaces*

Once the interfaces start to heat, the material begins to soften, even before sufficient heat is generated to promote flow. This heating can reduce the stiffness of the plastic, promote local deformation, and reduce the rate of particulate generation. As stated in Section 10.3.1, it is convenient to model the heating process through Eq. 10.1. By replacing the velocity component (Eq. 10.2) and force component (Eq. 10.3) in Eq. 10.1, it is possible to define heat generation in terms of welding parameters, as seen in Eq. 10.4.

$$P = F \cdot v$$
$$P = (F_n \mu) \cdot (r\omega) \tag{10.4}$$

Thus, heat generation is directly proportional to the clamp pressure, the controlling parameter of F_n, and the machine setting of rotational speed (ω). It should be noted that heat generation is defined as seen in Eq. 10.4 only during phase I of the welding process.

10.3.3   Viscoelastic Heating (Phase III)

As stated earlier, Phase II, the transition phase, is comprised of a mix of solid/solid frictional heating (Phase I) and viscoelastic shear heating (Phase III). Because this phase is a mix between two heating processes, it is a complex phase and it is beyond the scope of the book to examine it in significant detail.

Once Phase III starts, indicated by a relatively linear gradient between melt displacement and time, there is no solid/solid frictional heating. Instead, heat generation is dominated by shearing. In Phase III, the process can be modeled by two solid interfaces separated by a molten layer with a thickness of 2 h, see Figure 10.5.

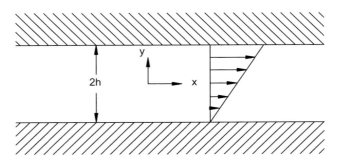

Figure 10.5 *Cross section of viscoelastic melt in spin welding joint*

Based on squeeze flow, it is possible to estimate the force required to shear the two interfaces past each other using Eq. 10.5. The first parameter is the viscosity of the melt, η. This is a measure of how easy the melt flows and it is a reciprocal function to the melt flow index. In addition, viscosity is complicated by the fact that it is dependent on the temperature of the melt and its shear rate, τ. Thus, this parameter is very complicated to model, even with a FEA technique. The other parameter effecting the heat generation in Phase III is the melt thickness (2 h) [8].

$$F = \frac{\eta V}{2h} bd \qquad\qquad (10.5)$$

η = melt viscosity
b = width of weld interface
d = depth of weld interface

With the resistance force defined, it is possible to use Eq. 10.4 to estimate the power generated during Phase III, as shown in Eq. 10.6.

It is interesting to note that there is a significant difference in the relationship between the process parameters and the amount of heat generated between Phase I and Phase III. For example, in Phase III heat generation is proportional to rotation speed squared while it is directly proportional to rotation speed in Phase I. In addition, during Phase I, heating is directly proportional to the clamp force, while during Phase III, there is no direct relationship between clamp force and heat generation. However, further examination of Eq. 10.7 does show that there is a direct relationship between heat generation and viscosity and melt thickness, which are both affected by clamp force. For example, with all other parameters equal, melt thickness is inversely proportional to clamp force because higher clamp force promotes squeeze flow of the melt out of the weld zone. Thus, although clamp force is not directly related to heat flow, it does promote thinner bond lines (small 2 h) and heat generation during Phase III. This relationship is further complicated because viscosity is related to clamp force. As noted earlier, the melt thickness is inversely proportional to clamp force and because of shear thinning, the viscosity of the melt is proportional to the melt thickness. This results in a self-limiting process. As the bond line becomes thinner, because of increased of clamp force or rotation speed, shear thinning reduces the melt viscosity and efficiency of heat generation. Thus, the bond line temperature is limited.

$$P = (F \cdot v)$$

$$P = (\frac{\eta v}{2h} bd)(v) = (\frac{\eta r \omega}{2h} bd)(r \omega)$$

$$\therefore$$

The heat flux generated at the interface (Q) is :

$$Q = \frac{P}{bd} = \frac{\eta r^2 \omega^2}{2h} \tag{10.6}$$

While shear thinning of the melt can limit bond line heating and prevent thermal run-away of the bond line, it can promote molecular alignment parallel to the bond line. This molecular alignment can result in relatively high residual stresses and increases the weld's sensitivity to solvent cracking. In addition, this molecular alignment can cause a loss of weld strength.

10.4   Equipment Description

10.4.1   Equipment Components

Spin welding systems vary greatly in design and complexity depending on their mode of operation (inertial drive system or servo-based drive system, see Section 10.2.2). However, there are several components common among most machines:

Generator/Power supply: This component converts line voltage into the correct voltage and current to power the controls circuits and servo motors if applicable. In pneumatic drive machines, a pneumatic motor supplies the spinning motion. Here care must be taken to provide proper air flow to the machine in order to assure equipment performance consistency. Air motors require the use of lubricated air, therefore care must be taken to prevent exhaust air from contacting the part, if a part is sensitive to contamination. Usually the equipment manufacturer will provide details in reference to airflow requirements. Servo-driven machine designs may still incorporate pneumatics for head delivery and weld force/clamp force application.

Controls: The controls establish an interface between the machine and the operator and monitor the system. They usually contain a PLC or similar type of logic circuitry. They interface the logic circuits to the user. The controller also reports to the operator on the conditions of the machine, such as welding data and machine status. The controls will vary greatly depending on mode of operation and machine design. For example, if the machine design is based on servo-drive, the operator may be required to input the angle of rotation or the number of revolutions.

Actuator: This component is usually a pneumatic press that moves the upper fixture (spin head) to the lower fixture and applies force during the weld and hold cycle. There is usually a pressure regulator, which determines the maximum force applied to the press. Some machine designs incorporate a servo-based actuator to generate the movement and force. Often, the actuator will incorporate an encoder in order to measure and report the amount of collapse. Figure 10.6 shows an example of a spin welder with a servo actuator for the z-axis. This type of system can control the orientation of parts at the end of the weld cycle. It is also possible to program such a system to rotate between two angles and not in a complete cycle, which is useful when the application does not allow full rotation between the two parts.

Figure 10.6 Spin welding actuator with servo control for spin and z-axis displacement *(Courtesy Branson Ultrasonics Corp.)*

Lower fixture: In many cases the lower fixture is a simple jig, which locates the parts being welded. If the machine design is based on a servo-drive where alignment can be critical to the application, care is often required to assure the fixture provides proper alignment, fit-up and is properly located. Usually there is a design with the fixture that accepts a mating feature in the part design to lock/couple the head to the part being spun. If the part does not have any features that can be grabbed, a manual or pneumatic clamping fixture can be used. This helps assure there is no relative motion between the part and fixture that may promote part marking.

Upper fixture: This component couples the drive mechanism to the part of the application being rotated. The upper fixture is often referred to as the drive head. Usually the head is designed with features that engage a mating feature in the part design to lock/couple the head to the part being spun. This reduces relative motion between the part and fixture that may promote part marking. In addition, the head may have counter weights added to the design to assure that the center of gravity (center of inertial mass) coincides with the center of rotation to reduce machine vibrations and side loads. This maybe needed in applications where the part is not symmetrical or it mass is not evenly distributed about the axis of spin during welding.

If there are no geometric features that can be used for drive engagement then a friction drive or an automatic clamping spin tool is necessary.

If the contact surface of the part is cosmetic and part marking is critical, a friction drive spin tool can use a soft coating like silicone rubber. If a soft-coated tool is used for production, it is recommended that a spare tool be purchased, because these tools will wear and require periodic re-coating.

If the contact surface is not cosmetic and part marking is allowed, then a more aggressive friction drive can be used. This can be achieved by a radial serration or knurl pattern, or with the use of abrasive coatings.

Before designing the upper spinning tool or drive features in the part it is necessary to determine how the parts to be assembled will be loaded in the spin welding machine. For example, if one part is in the lower fixture, and the other part is loaded into the drive tool, there is a lot of flexibility in the drive features used. If the part has any proud or recessed geometry, the upper tool can be designed to use these as drive features. Care must be taken to make sure that delicate features are not used, as they may be damaged.

Holding the part in the upper tool can be accomplished with the use of spring plungers, or with vacuum. Spring plungers are less expensive to use and are recommended as long as the geometry of the part is not prohibitive. For other parts, the use of rotary seals in the spindle block allows for the porting of vacuum to the spin tool. When vacuum is used, some filter media should be placed between the spin tool and the spindle. This prevents particulate generated during welding from damaging the rotary seals.

If both parts are loaded in the spin welding machine in the lower tool, care must be taken to determine how the upper tool will align with the drive features of the parts. As long as a friction drive tool is used this issue is usually not critical. For all other drive types, alignment of the upper tool must be addressed, such as by the use of alignment features within the lower fixture and upper part. This can be as simple as marking indicators for proper alignment between the upper part and the fixture.

If a servo-controlled spin welder is used, the radial position of the spin tool is known at the start of the cycle. If it is possible to control the radial orientation of the geometric features of the part to be contacted, then the spin tool can descend and engage those features. Controlling the radial orientation of the part can be accomplished with small break-away pre-alignment pins molded into the two parts to be welded or with precise placement by an automated system. There are also spin welding machines available that use the position information from a vision system to orient the spin tool to the starting position of the parts presented to it.

If there is design flexibility of the parts, the preferred technique is a self-locating drive feature, which has multiple drive ramps tapered to run into each other in the direction of spin rotation. The number of drive features can vary from 2 to 20 depending on the part design. This style of drive feature readily lends itself to automation. Figure 10.7 shows an example of self-locating drive features.

Figure 10.7 *Application with self-locating drive features*

10.5   Applications

There is a large variety of spin welding applications. Many assemblies that would not have been viable in the past are achievable with new spin welding techniques and equipment that has become available. Below are some sample applications that use spin welding techniques.

Relatively large (100 mm diameter) truck lights (Fig. 10.8) are a very common lighting assembly produced by a number of manufacturers. These lights are molded from polycarbonate, acrylic or ABS. They require a strong, leak-free weld and are typically leak tested after assembly. This style of light has been welded successfully with both a tongue and groove and with a shear joint design. The application works best when drive features are added to the lens, but they have also been welded using rubber friction drive tools. The assembly welds in approximately 2 to 4 revolutions.

Thermal drinking cups (Fig. 10.9) are typically made out of two components, an inner and an outer cup. In this application, the two parts are joined to form a cup with a sealed inner chamber to act as an insulator. The weld must be leak free to prevent the chamber from filling with water during washing/use, and flash must be controlled for a smooth and cosmetically appealing joint. The cups are commonly made in either acrylic or polypropylene. A tapered tongue and groove design is a typical joint design in this application (more detail in Section 10.5.1).

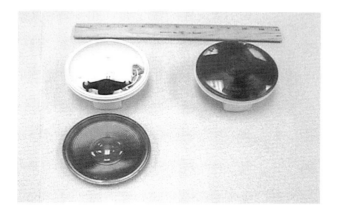

Figure 10.8 Truck lights with a 100 mm diameter (Courtesy Branson Ultrasonic Corp.)

Figure 10.9 Thermal drinking cup welded with spin welding (Courtesy Branson Ultrasonic Corp.)

Even non-classical spin welding applications work well in production. For example, the plumbing fitting shown in Figure 10.10 features a classical spin weld with a round part to the housing, and a non-symmetrical elbow to the housing. When spin welding, it is important to keep in mind that it is only necessary to have a round part *interface*, not a round assembly. If radial orientation of the final assembly is required, a servo driven spin welder is necessary.

Figure 10.10 *Non-circular application welded with spin welding (Courtesy Branson Ultrasonic Corp.)*

Spin welding is a joining process that promotes significant mixing at the faying surface, and due to this, it can often be used to join materials that are typically not compatible or weldable. For example, in Figure 10.11 a disc cut from rubberized cork is attached to an acetal gear with spin welding. While not all material combinations can be welded with spin welding, it is recommended testing prototype parts before ruling out a given combination.

Figure 10.11 *Rubberized cork-to-acetal gear welded with spin welding (Courtesy Branson Ultra-sonic Corp.)*

10.5.1   Joint Design

As with all plastic joining processes it is important to have the correct joint design to achieve the best results. Spin welding is very aggressive and flash and particulate are usually generated during the welding process. It is important to make sure that the part design considers this. The proper use of flash traps will contain most of the displaced material. Spin welding works with many styles of joint design and some of the most common designs are detailed below.

The tongue and groove (Fig. 10.12) design with a 30° included angle is a very strong weld joint and provides some self-alignment characteristics. Pressure and filter vessels using this design typically fail outside of the weld joint during burst tests. The natural lead-in of the taper-to-taper fit helps to successfully weld parts that are slightly out of round, or have large tolerances. The weld collapse is approximately 0.69–0.76 mm.

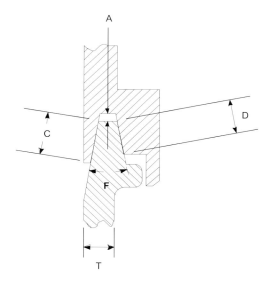

Figure 10.12 *Tapered tongue and groove with 30° included angle. T = Wall thickness, A = ~0.5 T to 0.8 T, F = 30°, C + D = Weld surface (~2.5 x T) (Courtesy Branson Ultrasonic Corp.)*

The shear joint design (Fig. 10.13) is commonly used for high strength welds in thin-walled parts or parts with too little real estate for a tapered tongue and groove. For maximum strength, the weld collapse should be at least 1.5 times the wall thickness. It is important that the interference in the weld joint stays consistent. The flanged shear joint (Fig. 10.14) can help maintain interference, and aid in flash containment.

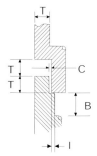

Figure 10.13 *Details of typical shear joint. T = Wall thickness, B = 1.5 x T, C = Clearance (~0.13mm), I = interference (~0.25 to 0.50 mm per side) (Courtesy Branson Ultrasonic Corp.)*

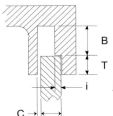

Figure 10.14 Details of typical flanged shear joint. T = Wall thickness, B = 1.5 x T, C = Clearance (~0.13 mm), I = interference ⁻0.51 to 1.1 mm per side)

When welding assemblies molded in nylon, it is common in industry to use a modified shear joint. The nylon shear joint (see Fig. 10.15) increases interference with the depth of the weld.

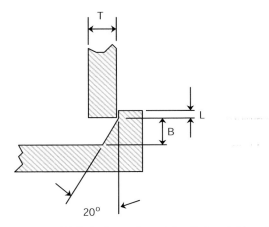

Figure 10.15 Typical shear joint. T = Wall thickness, B = 1.5 x T, L = 0.51 to 1.1 mm (Courtesy Branson Ultrasonic Corp.)

10.6   Material Weldability

Most thermoplastic materials can be joined with the spin welding process. As detailed in Section 10.3, a material with a relatively high coefficient of friction is critical to promote heating and melting. Thus, materials such as fluorpolymers with a relatively low coefficient of friction are typically not joined with the spin welding process. In addition, it is generally best if the materials being joined are compatible. As detailed in Chapter 2, there are many material characteristics determining material compatibility, such as morphology, melt flow, and melt/flow temperature. Because of the complexity of these characteristics and their interactions, it is recommended that compatibility (weldability) tests be performed on particular materials to determine their weldability. Table 10.3. provides general weldability guidelines for selected materials with the spin welding process.

Table 10.3 *Typical Spin Weldability of Selected Polymers [9]*

MATERIAL	WELDABILITY	COMMENTS
ABS	Good to Excellent	
ABS/polycarbonate alloy	Good	
Acetal	Fair to Good	Over-welding can weaken joint
Acrylic	Good	
Acrylic multi-polymer	Good	
Acrylic styrene acrylonitrile	Good	
Amorphous polyethylene terephthalate	Poor to fair	
Butadiene styrene	Good to excellent	
Cellulosics	Good	
Polyvinylidene fluoride (PVDF)	Good	
Perfluoro alkoxy alkane (PFA)	Poor	
Liquid crystal polymers	Fair to good	
Nylon	Good	
PBT/polycarbonate alloy	Good	
Polyamide-imide	Fair to good	
Polyarylate	Good	
Polyaryl sulfone	Good	
Polybutylene	Poor to fair	
Polybutylene terephthalate (PBT)	Good	
Polycarbonate	Good to excellent	

Table 10.3 *Typical Spin Weldability of Selected Polymers [9] (Continuation)*

MATERIAL	WELDABILITY	COMMENTS
Polyethylene terephthalate (PET)	Fair to good	
Polyetheretherketone (PEEK)	Fair	Generates fine particulate
Polyetherimide	Good	
Polyethersulfone	Good to excellent	
Polythylene	Good	Generates a lot of particulate
Polymethylpentene	Good	
Polyphenylene oxide	Good	
Polyphenylene sulfide	Good	
Polypropylene	Good to excellent	
Polystyrene	Good to excellent	
Polysulfone	Good	
Polyurethane	Poor to fair	
PVC (rigid)	Good	Fumes must be vented
Styrene acrylonitrile	Good to excellent	

It is also important to note that because the spin welding process promotes significant mixing at the faying surface due to the spinning motion, it is sometimes possible to join materials that are not compatible. This type of bond is detailed in Section 10.5 (Fig. 10.11). In this case, the two materials are joined by mechanical interlocking of the polymer into the cork. That is to say, that is possible to force the molten polymer into a solid porous substrate, cork in this example, and allow the two materials to interlock together to form a joint. It is important to note that while this is not a conventional weld, the resulting joints can be relatively strong.

10.7   Setup and Troubleshooting

As with any of the welding processes, trouble shooting an application often requires the assistance of a person that is properly trained, such as an application engineer or technician from the equipment manufacturer. However, Table 10.4, provides guidelines for typical problems with associated possible causes and solutions.

Table 10.4 *List of Selected Problems and Their Possible Solution for Spin Welding*

PROBLEM	CAUSE	SOLUTION
Vibration while spinning	Spin tool, or spin tool and part combination are out of balance	Balance the spin tool. If the problem persists, balance the spin tool with the plastic part that is loaded into the spin tool installed.
Part marking	Slippage in the fixture	This is most common when a friction drive tool is used. Friction drive tools will wear over time and start to slip while welding. The tool must be re-worked if made out of steel, or re-coated if silicone rubber is used. A more permanent fix is to add drive features to the parts.
Excessive flash	The assembly is over-welded	Reducing the number of revolutions of weld or weld pressure.
Excessive/significant flash, but weak weld	Lack of fusion at the joint line	Make sure that the spin welder has rapid deceleration or active braking. If the spin tool coasts to a stop it is likely to tear bonds that are being formed as the tool slows down and the material cools. It is also possible too excessive clamp pressure prevents proper fusion, thus a reduced clamp pressure may reduce this problem.
Excess particulate	Lack of proper melting	Increasespin RPM or weld force. If the spin RPM or clamp pressure is too low, the abrasion phase of the spin weld may be extended causing excess particulate gener-ation. Increasing one or both can decrease the time to enter the melt phase.

10.8   References

1. Tappe, P., Potente, H., *New Findings in the Spin Welding of Plastics*, ANTEC-89, Conference Proceedings, Society of Plastics Engineers, Brookfield, CT

2. Branson Ultrasonics Corp., *Powered Spin Welder Model SW200-G*, Data Sheet 1999, Branson Ultrasonics Corp. Danbury, CT

3. Stokes, V.K., *Vibration Welding of Plastics. Part I: Phenomenology of the Welding Process*, Polymer Engineering and Science, Vol. **28**, No. 11, June 1988

4. Stokes, V.K., *Vibration Welding of Plastics. Part II: Analysis of the Welding Process*, Polymer Engineering and Science, Vol. **28**, No. 11, June 1988

5. Ehrenstein, G.W., Giese, M., *Identification of Phase 3 in the Vibration Welding Process*, ANTEC-93, Conference Proceedings, Society of Plastics Engineers, Brookfield, CT

6. Sear, F.W, Zemansky, M.W., Young, H.D., *University Physics*, 6th Edition (1982) p. 25–33, Addison-Wesley Publishing Company, Menlo Park, CA

7. Steijn, R.P., *Friction Wear of Plastics*, Met. Eng. Quart., ASM, May 1967

8. Stokes. V.K., *Analysis of Friction (spin)-Welding Process for Thermoplastics*, Journal of Materials Science, Vol. **23** (1988) p. 2772–2785, Chapman and Hall Ltd.,

9. Rotheiser, J., *Joining of Plastics*, (1999) p. 418, Hanser Publishers, Munich

11 Radio Frequency Welding

James P. Dixon and David Grewell

11.1 Introduction

Radio Frequency (RF) welding, which is also often referred to as "dielectric welding" is a process that relies on internal heat generation by dielectric hysteresis losses of thermoplastics. It is most commonly used to weld PVC bladders such as intravenous drip bags for the medical industry. It is also used to weld book and binding covers as well as blister packages.

RF welding has the advantages that it is a relatively fast process with typical cycle times of less than 2 to 5 s. It also does not require any special joint designs and produces relatively appealing welds.

RF welding is almost exclusively used for welding relatively thin sheets or films. Thickness' usually range from 0.03 to 1.27 mm, depending on the material and the application. The welding of films is limited because a strong electric field must be generated and this can only be achieved when the welding electrodes are brought together in close proximity, 0.03 to 1.27 mm. If the welding electrodes are significantly further apart, the electric field density is too low to effectively heat and melt the plastic.

Another limitation of the process is that the material being joined must have the proper electrical properties. One such property is a relatively high dielectric constant, typically greater than two. This allows more current to flow through the material, which promotes heating, at a lower electrode voltage.

The major restriction on materials to be welded by RF is the fact that the material must have a relatively high dielectric loss, i.e., the material must have the ability to convert the alternating electric field into heat. The most common thermoplastics that can be effectively heated with RF welding are:

- PVC
- CPVC
- Polyurethane
- Nylons (polyamide)
- Cellulose acetate
- PET, PETG

A relatively high dielectric breakdown voltage (dielectric strength) in the material is also important to reduce the possibility of arcing through the material. This property is usually less of a restriction for material selection with RF welding, because most plastics are good electrical insulators and inherently have high dielectric breakdown voltage. In some cases, a third material can be added to the bond line to allow for materials, which naturally cannot be heated by RF to be welded.

11.2   Process Description

11.2.1   RF Welding

There are five major steps in the RF process:

1) *Loading parts.* This usually involves laying sheets of the materials to be welded onto the lower fixture and is not considered a critical operation.

2) *Press actuation.* This step is initiated by the operator and is usually accomplished by the closure of two palm buttons. If the machine is automated, the fixture will translate the material to be welded into the press system. The press will then lower the upper platen/electrode so that the sheets are clamped between the electrodes by a pre-selected clamp force.

3) *RF application.* Once the sheets are clamped, the RF electric field is applied. This promotes heating and melting of the material clamped by the electrodes. The heating time is usually pre-selected by the operator.

4) *Hold phase.* After the RF energy is applied to the sheets, the press maintains pressure for a pre-selected length of time (called the hold time) to allow the melt to solidify under pressure.

5) *Part unloading.* If the equipment is automated, the press will open the electrodes and the lower fixture will translate out of the press unit and allow the operator to remove the welded sample(s).

Table 11.1 contains a listing of the typical times required for the five RF welding steps.

Table 11.1 Typical RF Welding Cycle

Step	Typical time
Loading parts	3 to 15 s
Press actuation	1 to 2 s

Step	Typical time
RF application	1 to 5 s
Hold phase	1 to 3 s
Part unloading	3 to 5 s (additional time with tear seal)

11.2.2   RF Sealing and Cutting

The simultaneous cutting and welding of the part involves the same steps reviewed in the previous section. The electrode configuration is the only major difference (see Fig. 11.1). Because the electrodes, even at the cutting edge, cannot be allowed to make contact while the electric field is applied, there is usually a small amount of material remaining at the toe of the weld. The operator must tear the samples along this section of the weld. Thus, it is common to refer to this type of operation as a tear seal and not a cut.

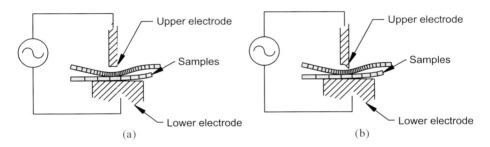

Figure 11.1 *Typical electrode configuration for (a) welding and (b) cutting and sealing*

If the electrodes are allowed to make contact and truly cut the samples, two problems can occur:

1. Power supply damage due to arcing
2. Excessive wear of the electrodes

It is possible to use a backing sheet to help minimize the need of retaining a small amount of material at the tear zone, but this sheet may need to be replaced frequently, depending on the application. Common materials that are used as backing sheet include phenolic, mylar® or Teflon®.

11.3   Physics of the Process

11.3.1   The Electric Field

In RF welding, an electric field is generated by charging two plates with opposing electric potentials. The voltage of the potential is both application- and material-dependent and is usually set during the tuning process of the equipment setup (see Section 11.7). The typical voltage potential, e_p, ranges from 1,000 to 1,500 VAC, at a frequency (ω) of 27.12 MHz.

In order to understand the electric field that is generated between the two electrodes, a simplified model of two plates will be used to represent a typical welding configuration. The model seen in Fig. 11.2 shows four primary variables:

- Frequency (ω)
- Voltage potential (v_p)
- Electrode separation (d_g)
- Relative dielectric constant of gap material (ε_r)
- Dielectric constant in a vacuum (ε_o)
- Electric field strength ($E = v_p/d_g$)

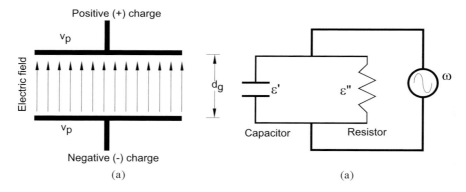

Figure 11.2 (a) Model of electric field between two plates and (b) Electrical model of dielectric welding

While there are several possible electrode configurations, the most common one is a lower electrode that is best described as a flat plate with the opposing (upper) electrode being comprised of a copper or brass sheet metal on edge (1 to 3 mm thick), see Fig. 11.3.

Another, less popular, configuration consist of two opposing electrodes, both comprised of copper or brass sheet metal on edge (see Fig. 11.4). This is less popular, since electrode to electrode alignment is critical in order to assure proper welding. However, some applications, such as some book covers, require this configuration in order to accommodate the part geometry.

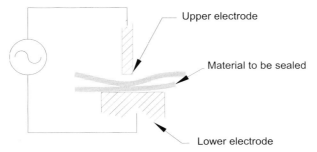

Figure 11.3 *Typical electrode configuration, flat plate opposing sheet metal on edge*

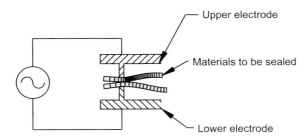

Figure 11.4 *Typical electrode configuration, two opposing sheet metal electrode*

While the electrode configurations produce different electric field contours (Fig. 11.3 and Fig. 11.4), the model, a capacitor, as seen in Fig. 11.2 can be used to estimate heat generation [1]. The capacitance (C_o) of the model is:

$$C_o = \frac{\varepsilon_o \varepsilon_r A}{d_g} \tag{11.1}$$

where ε_o is the permittivity of a vacuum, ε_r is the permittivity of the material (plastic), and A is the cross sectional area of the weld.

It is important to note that compared to typical capacitor models, the model seen in the figure includes a lossy component (resistor). To model the dielectric welding process, where the losses are substantial, the resistor must incorporated into the model. That is to say, the loss (resistor component-ε') and storage (capacitance component-ε'') components of the dielectric constant must be incorporator. The complex loss modulus is the vector sum of the two components and is defined as:

$$\varepsilon_r^* = \varepsilon' - i\varepsilon'' \tag{11.2}$$

Thus the complex capacitance (C^*, loss and storage components) is defined by substituting Eq 11.2 into 11.1; giving:

$$C^* = \varepsilon_r^* C_o = (\varepsilon'-i\varepsilon'')C_o \tag{11.3}$$

By modeling the plastic between the electrodes as a resistor (loss component) and capacitor (storage component) in parallel, see Fig. 11.2, it is possible to derive the voltage and current. If it is assumed that that the voltage is defined as:

$$v_p(t) = V_0 \sin(\omega t) = V_o e^{i\omega t} \tag{11.4}$$

where $v_p(t)$ is the instantaneous voltage and V_o is the peak voltage. It is important to note that in this case the frequency (ω) is in units of radians/s.

The instantaneous current $(I(t))$ through a capacitor is defined as:

$$I(t) = \frac{dq}{dt} \tag{11.5}$$

where dq/dt is the rate change of charge. Since the charge $q(t)$ on a capacitor is defined by:

$$q(t) = C * v_p(t) = (C_o \varepsilon'-i C_o \varepsilon'')\left(V_o e^{i\omega t}\right)$$

$\therefore$

$$\frac{dq}{dt} = \underbrace{\left(i C_o \omega \varepsilon' V_o e^{i\omega t}\right)}_{storage} + \underbrace{\left(C_o \omega \varepsilon'' V_o e^{i\omega t}\right)}_{loss} \tag{11.6}$$

By only considering the loss component of the current (dq/dt) and putting in terms of a sine function it is possible to determine the instantaneous power $P(t)$ by multiplying the voltage (Eq. 11.4) and current (Eq. 11.6-loss):

$$P(t) = (C_o \omega \varepsilon'' V_o e^{i\omega t})(V_o e^{i\omega t})$$

$$P(t) = \left(C_o \omega \varepsilon'' V_o \sin(\omega t)\right)\left(V_0 \sin(\omega t)\right) = C_o \omega \varepsilon'' V_o^2 \sin^2(\omega t) \tag{11.7}$$

To determine the average power, P_{avg}, the instantaneous power (Eq. 11.6) must be integrated over a cycle:

$$P_{avg} = \frac{\omega}{2\pi} \int_{0}^{2\pi/\omega} C_o \omega \varepsilon'' I_o'^2 \sin^2(\omega) dt = \omega \varepsilon'' C_o I_{rms}'^2 \qquad (11.8)$$

The average volumetric heating (Q_{avg}) is defined as:

$$Q_{avg} = \omega \varepsilon'' \varepsilon_o E_{rms}^2 \qquad (11.9)$$

11.3.2   Molecular Heating

In many chemical structures, the electrical charge is unevenly distributed, resulting in a diatomic molecule [2]. The amount of uneven charge distribution is measured in units of debye, D, where one D is equivalent to 3.33×10^{-30} coul-m [3]. This measure is referred to as the dipole moment. A debye is proportional to the relative charges (Q) and the distance separation (r) of charges (Eq. 11.6). Thus, a molecule with larger relative charge or charge separation will have more dipole moment.

$$D = Qr \qquad (11.6)$$

Among the many chemical structures that produce a dipole, one of the strongest is the one in a salt. For example, in standard table salt, NaCl, the chlorine atom has a much higher affinity for the electrons compared to the sodium atom, thus the electron cloud (negative charge) tends to be distributed more around the chlorine atom. This leaves the sodium atom with a relative positive charge (see Fig. 11.5(a)). However, if the molecule is balanced, it may not have a dipole, even though the electrons are not evenly distributed, such as in the case of carbon dioxide (CO_2) (see Fig. 11.5(b))

(a)   (b)

Figure 11.5 Schematic of electron distribution of: (a) NaCl and (b) CO_2 molecules

In other polar molecules, the uneven distribution is generated because of asymmetrical structures. For example, in the molecule (H_2O) the molecule is bent at 104.5^0 due to hydrogen bonding. In the case of polar polymers, the uneven distribution is typically generated by side-groups that have a relatively high electronegativity. These include side groups such as chlorine and oxygen, which have an electronegativity of 3.2 and 3.4, respectively.

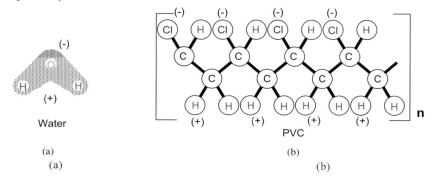

Water
(a)
(a)

PVC
(b)
(b)

Figure 11.6 Schematic of: (a) water and (b) PVC molecules with polar structure

It is also possible to induce a polar moment into a non-polar molecule with an electric field, referred to as polarization. However, this type of dipole only exists while the electric field is applied and it is not an effective method to heat and weld plastics. Placing a dipole molecule into an alternating electric field can be a very effective way of heating the material. During each alternating cycle, the molecule tends to align itself with the field. In some cases, such as with water, the entire molecule changes orientation (see Fig. 11.7). In other cases, such as in large polymer molecules, the alignment is locally strained at the dipole charges.

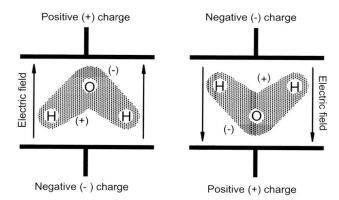

Figure 11.7 Example of molecular alignment of water between two charged plates of an alternating field

This rapid change in molecular orientation translates into molecular movement and vibrations. As the molecules vibrate, their motion tends to lag the phase of the electric field due to molecular friction and inertial effects. This difference in phase is defined as the loss angle (δ) [4]. Thus, with increased molecular vibrations due to alternating molecular orientation the material in the field begins to heat.

Beyond the electric field strength and frequency, heating is related to:

*   Molecular dipole strength/dipole moment
*   Molecular freedom

It is important to understand that, as with many other loss properties of plastics, the dielectric loss is temperature- and frequency-dependent. For example, as the material heats, the free volume in the sample increases allowing increased molecular movement for a given applied electric field. Thus, in most plastics the dielectric loss is proportional to temperature. This must be considered during equipment setup, because the close relationship can result in thermal runaway (rapid temperature rise).

11.4   Equipment Description

11.4.1   Equipment Components

An RF welding system consists of six major components:

Generator/Power Supply: This component converts line frequency into high frequency (27.12 Mhz) high voltage, by rectifying the AC line voltage to DC for the DC bias and supplying current to the vacuum tube. The voltage is selected by the operator and is application-dependent. The generator usually contains an oscillator to generate the desired frequency. In addition, the generator usually has power, tuning, and overload circuits. A more detailed description of the generator can be found in Section 11.7.

Controls: The controls interface the machine with the operator and monitor the system. They usually contain a PLC or similar type of logic circuitry. The most popular mode of welding with RF is the time mode, in which case the operator sets the controller to heat the parts for a predetermined length of time. The controller also reports specific conditions of the machine, such as power dissipation and tuning, to the operator.

Actuator: This component is usually a pneumatic press that moves the upper fixture (platen) to the lower fixture and applies force during the weld and hold cycle. There is usually a pressure regulator which determines the maximum force applied to the press.

Lower Fixture: In many cases the lower fixture is simply a large flat plate. It acts as a lower support for the parts being welded and for the lower electrode to apply the electric field. In some cases, the lower fixture contains raised electrodes to allow parts with tall dimensions to be welded. In automated systems, the lower fixture can translate the part in and out of the machine. This translation is accomplished either with a rotary or with a slide table.

Upper Fixture: The upper fixture is often designed and built by the end user. The fixture applies the electric field and the localized clamp force to assure proper welding. The electrodes are typically fabricated from 2 to 4 mm thick copper, bronze, or brass sheet metal. These metals have good electrical and thermal conductivity; both properties enhance welding. They are also easy to machine. Because it is common for many processors to fabricate their own fixtures, additional information is provided in Section 11.4.2.

RF Enclosure (Faraday Cage):   In higher power equipment (+5 KW), a cage is placed around the tooling/electrodes to protect the processor from the high voltage of the electrodes as well as any RF radiation hazards. The hazards of RF radiation were studied in detail and the reader is advised to refer to the IEEE (The Institute of Electrical and Electronics Engineers) C9.51-1999 [5] as well as to the CFRD (Center for Devices and Radiological Health [6]) regulations for additional information and references. The cage is usually fabricated from a metal screen to allow proper shielding without restricting viewing of the tooling and process.

Figure 11.8 shows a photograph of a typical RF machine and identifies all components described above.

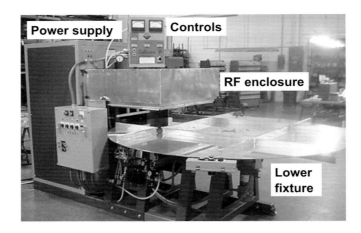

Figure 11.8 *Typical RF welding system (Courtesy Dielectrics)*

11.4.2   Fixture Design

Because each application requires a separate dedicated fixture, it is common practice for the fixture to be fabricated by the end user. In the simplest tools, the upper tool is a raised projection of metal that matches the geometry of the parts being welded and the lower fixture is a flat metal plate (see Fig. 11.9). This type of fixture is easy to fabricate and maintain, but this has the two main disadvantages:

1. Fringing of the electric field that produce melting beyond the upper electrode area. This is similar to the effect of a magnetic field extending outside the area of the magnetic pole.

2. Application must be able to accommodate flat lower fixture.

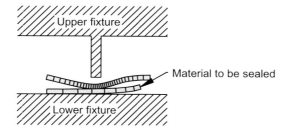

Figure 11.9 Cross-section of typical flat–to-raised fixture design

In more critical applications, where cosmetics are important and/or application shape does not allow for a flat lower fixture, the lower fixture may also have a raised electrode protrusion as seen in Fig. 11.10. This type of fixture defines the electric field better and produces less fringing and thus results in a more defined weld area compared to a raised/flat electrode. However, with this configuration, alignment is critical.

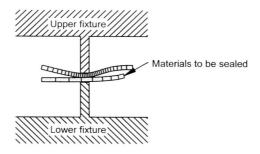

Figure 11.10 Cross-section of raised-to-raised fixture

In fabricating a fixture it is possible to machine the raised electrodes from a solid plate of metal, however, it is more common to bend the electrodes from a standard electrode profile. Commercially available profiles are usually made of brass or copper to maximize electrical and thermal conductivity. In addition, these materials are easy to machine. The profiles are pre-drilled for attachment to the lower plate and offer various surface finishes:

• Flat
• Knurled
• Cut and seal profile
• Stitched

Figure 11.11 shows a photograph of commercially available profiles [7]. It is important to note that while many of the profiles have complex geometry, they do not have any sharp points that can concentrate the electric field and promote arcing or localized heating.

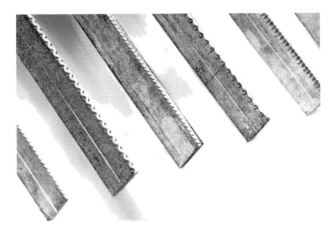

Figure 11.11 Photograph of a typical profiles used for fixtures fabrication

It is also common for the lower fixture (flat plate portion) to be covered with a non-stick, high-dielectric film. This helps release the parts from the fixture, increases equipment efficiency and power delivery. Common materials include:

• Fish paper
• Kapton films
• Teflon films

In most cases these films must be replaced on a periodic basis due to thermal and/or mechanical wear.

In applications that have high duty cycles or high production rates the fixture can be water-cooled to prevent heat build-up and to minimize cycle times. When welding crystalline materials, which have a relatively distinctive melt/processing temperature, the fixtures are sometimes slightly heated with an electrical heater to minimize cycle times.

11.4.3   Selection of the Proper Equipment and Power Requirements

There are two main considerations for equipment selection:

• Automation requirements
• Power requirements

When considering the automation requirements, the main issue is manufacturing needs. If the cycle rates are low to moderate, i.e., 10 to 60 parts per hour, it may be sufficient to acquire a welding system with very little automation. However, with higher manufacturing

Table 11.2 *Material Weldability with RF Welding*

MATERIAL	WELDABILITY RANKING (1–5) 1 = Best result, 5 = Not Possible
ABS	2
Acetal	4
Acylics	3
Aclar	3
APET	2
Barex 210	1
Barex 218	1
Butyrate	2
Cellophane	5
Cellulose acetate	2
Cellulose nitrate	3
Cellulose triacetate	3
Ethyl cellulose	5
EVA (ethyl vinyl acetate)	2
EVOH (ethyl vinyl alcohol)	3
Nylon (PA)	2
Pelathane (not recommended w/tear-seal)	2
PET (polethylene teraphthalate)	2
PETG (polyethylene terephthalate glycol)	1
Phenol-formaldehyde	2
Pliofilm (rubber hydrochloride)	1

Table 11.2 *Material Weldability with RF Welding (Continuation)*

MATERIAL	WELDABILITY RANKING (1–5) 1 = Best result, 5 = Not Possible
Polyethylene	5
Polymethylacrilate	3
Polypropylene	5
Polycarbonate	4
Polystyrene	5
Polyurethane	2
Polyurethane foam	3
Polyviny acetate	2
Polyvinyl chloride, flexible & clear	1
Polyvinyl chloride, pigmented	1
Polyvinyl chloride, opaque	2
Polyvinyl chloride, semirigid	2
Polyvinyl chloride, rigid	3
Polyvinyl chloride, flexible w/glass-bonded	1
Polyvinyl chloride, coated paper or cloth	1
Rubber	5
Saran (Polyvinylidene chloride)	1
Silicone	5
Teflon (tetrafluoroethylene)	5

11.7   Setup and Troubleshooting

Setting up an RF welding application requires three basic steps:

*   Fixture alignment
*   Equipment tuning
*   Parameter adjustment

Fixture Alignment: Fixture alignment is usually the simplest of the three steps, and involves assuring that the electrodes apply uniform pressure to the parts. This process can become more complicated when two opposing raised electrodes are used as seen in Fig. 11.10. Using pressure sensitive paper (carbon paper) as a guide helps to overcome this difficulty. An easier approach involves using locator pins incorporated within the design of the fixtures to help with alignment. During alignment, the electrodes are brought together under the weld force (clamp force), but the RF energy is not applied. Often the electrodes are leveled using sheet metal shims to adjust their height in order to assure uniform pressure across the entire weld zone. After the fixtures are properly aligned and leveled, care should be taken to remove any sharp corners or edges from the fixtures to reduce the possibility of arcing.

Equipment Tuning: While there is a wide range of equipment designs on the market for RF welding, most are designed so that the equipment must be properly tuned to the load. This is similar to tuning a radio to a particular radio station. However, with RF welding, there is a slight drift in the tuning frequency, caused by several effects: as the weld progresses, the gap between the electrodes closes and in turn increases the electrical capacitance of the load. Also, as the material melts, the dielectric constant increases slightly, further increasing the capacitance of the load. As the capacitance of the load increases, the operating (tuned) frequency of the system tends to decrease. Thus, in a manually tuned system, the set point operating frequency is usually set below the resonant frequency of the system. Thus, when RF power is applied to the fixture, the system is slightly de-tuned below the optimum tuning point and as the weld progresses, the operating resonant (tuned) frequency decreases, closer matching the operating frequency. When the operating frequency matches the resonant frequency, the system is tuned and 100% of the available power will be delivered to the load. Again, this frequency will be near 27.12 MHz, so that the machine complies with FCC regulations.

There are several ways to adjust the resonant frequency of the system. While the resonant frequency can be greatly adjusted by making changes in the fixture, this is usually not feasible for a given application because part modifications will be required. Instead, most RF equipment has two tuning variables:

*   Coupling capacitance
*   Coupling inductance

Referring to Fig. 11.16, a simplified model of the electrical circuit of most RF welding systems, the system can be broken into four major components:

1. Load: The electrical equivalent to the load including the electrodes, fixtures and application

2. Cable: The electrical equivalent to the cables between the generator and fixture

3. Tuning: The electrical equivalent to the tuning circuit within the generator

4. Generator: The electrical equivalent to the generator that converts line power to high voltage high frequency

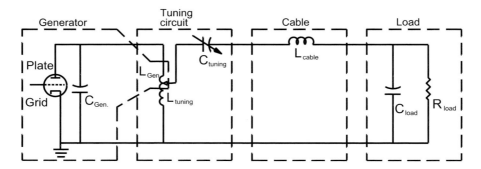

Figure 11.16 *Simplified electrical circuit of an RF welding machine*

While varying of any of the above components will change the resonant frequency of the system, the operator usually only varies the components within the tuning circuit. Before the main power is applied to the generator, it is usually advisable to determine the resonant frequency of the tuning circuit, cable and load, (as one unit), by shorting the generator and using a grid dip oscillator (GDO). If the frequency is below 27.12 MHz, the tuning capacitance (C_{tuning}) should be adjusted so as to increase the resonance frequency of the system (tuning circuit, cable and load). In most equipment, the tuning capacitance (C_{tuning}) can be adjusted by moving a slide on the control panel of the generator. If there is insufficient adjustment in the C_{tuning}, further tuning adjustment can be made by adjusting the tuning load (L_{tuning}). It is only advisable to adjust the L_{tuning} as a last resort because this circuit is shared with the generator and a change can affect the overall resonant frequency of the system. In some cases, such as very large or very small applications, it may be necessary to add additional capacitance or inductance to the system. If this is required, it should be completed only by trained personal or under consultation of the equipment manufacturer.

Once the C_{tuning} and L_{tuning} are properly set, the generator should be un-shorted and the main power supplied to the generator. During a weld cycle, the grid current should be monitored and the C_{tuning} should again be adjusted to minimize the grid current, a procedure referred to as "dipping the current."

If the equipment has an auto-tuning circuit, most of the above mentioned issues are irrelevant. The close-looped circuit in an auto-tuning system automatically adjusts the tuning in order to assure that maximum power is being delivered to the load. The system will even adjust during the weld cycle, so it is important to pre-select a maximum power output level to prevent overloading the machine or overheating the application.

Parameter Adjustment: Once the system is properly tuned, the final step in equipment set-up is adjusting the welding parameters. Table 11.3 contains a list of possible problems and corresponding adjustments.

Table 11.3 *Listing of Common Problems and Possible Solution for RF Welding*

Condition/problem	Possible solution
Over-welded	Reduce weld time
	Reduce power output
Under-welded	Increase weld time
	Increase power output
Too much flash	Reduce clamp pressure
	Reduce weld time
	Reduce power output
	Assure proper fixture alignment
Part marking	Assure proper fixture alignment
Arcing	Apply buffer sheet (fish paper, Teflon)
	Check for sharp corners on fixture
	Reduce weld time
	Reduce power output
Overload	Assure tuning, limit power output
Sticking to fixture	Apply buffer sheet (fish paper, Teflon)
	Reduce hold time
Over-heating of fixture	Apply buffer sheet (fish paper, Teflon)
	Add chiller to fixture plates

11.8   References

1. Faisst, C.F., Benatar, A., *A Feasibility Study of Radio Frequency Joining of HDPE Using Conductive Polymeric Composite,* ANTEC 1994 Conference Proceedings, Society of Plastic Engineers, Brookfield, CT

2. Brown T. L., Lemay, Jr. H. E., *Chemistry The Central Science* (1981) Prentice-Hall Inc., Englewood Cliffs, NJ

3. Sears, F.W., Zemansky, M.W., Young, H.D., *University Physics* (1982) Addison-Wesley Publishing Co., MA

4. Billmeyer, W.F. *Textbook of Polymer Science,* (1984) John Wiley and Sons, Inc., New York

5. The Institute of Electrical and Electronics Engineers, 445 Hoes Lane, PO Box 459, Piscataway, NJ 08855-0459

6. Center for Devices and Radiological Health, 2098 Gaither Road, Rockville, MD 20850

7. Watson, M.N., *Joining Plastics in Production,* (1988) Crampton & Sons Limited, Sawston, Cambridge, UK

8. "Introduction into RF Sealing for Clam Shell Blister Packages", Alloyd RF Sealing System, Technical Bulletin, January 1998

12 Infrared and Laser Welding

David Grewell

12.1 Introduction

Infrared (IR) and laser welding of plastics has been available for many years, but only recently, with the decrease in price for lasers and laser diodes as well as with a higher demand for part quality, has it become more popular. In addition, there has been significant research regarding material aspects, tooling design, and optics during the 1990s from which this technology has profited significantly.

When referring to IR/laser welding, it is not possible to describe it as a single joining technique. There are several modes of IR welding/joining. Most notably:

1. *Through Transmission IR Welding (TTIr)*.   The technique is based on passing IR/laser radiation through one component, which is IR transparent and having the second component, which is to be joined, absorb the radiation. An application that is well suited to TTIr is the joining of automotive taillights. The outer lens is transparent to IR radiation with wavelengths between 800 and 1000 nm, while the housing is often filled with a die or pigment, which is IR absorbent. There are three modes commonly used with TTIr:

   * *Plunge*
   * *Scanning*
   * *Masked scanning*

2. *Surface Heating*.   This technique is very similar to heated tool or hot plate welding. The surfaces of the components to be joined are heated by direct IR/laser exposure for a sufficient length of time to produce a molten layer, usually for 2–10 s. Once the surface is fully melted, the IR/laser tool is withdrawn from between the parts and the parts are forged together and allowed to solidify.

3. *IR/Laser Staking*.   This technique deforms a thermoplastic stud into a button geometry to produce a mechanical fastener similar to a rivet. This is usually accomplished by having a specially designed tool, which allows the IR/laser radiation to be directed onto the stud. After the stud is softened by IR radiation, it is deformed into a cavity that captures the molten plastic to produce the desired button geometry. In this case, the top surface of the cavity must be fabricated from a transparent material, such as glass, to

allow the IR to be directed onto the stud. In some cases, it is not possible to fabricate the tool with a transparent top surface, such as in applications containing a glass filler that would promote wear of the tool. In these cases, the stud is heated with IR radiation, then a separate tool is translated into position to capture and deform the melt.

There are many sources of IR radiation, including lasers, quartz lamps, and ceramic heaters. The emitted radiation is specific for each source and is often the determining factor for selecting the source. For example, with TTIr welding laser sources, laser diodes and YAG lasers are often used because they produce mono-chromatic (narrow band width) light and can be well focused.

There are many advantages for using IR/laser welding, some of which are listed below:

• Fast cycle time, typically 2 to 10 s.
• No part marking from tooling and fixture
• No particulate generated
• Flash is smooth and fully attached to part (not free to break away)
• Relatively low residual stresses
• Heat affected zone can be well defined

Some of the limitations of IR/Laser welding include:

• Capital costs can be relatively high
• Some materials are not well suited, such has highly filled crystalline materials
• Some applications' geometry is not well suited, e.g., applications with internal walls that cannot be exposed to the IR/laser radiation.

12.2   Process Description

There are three basic modes or techniques of IR/laser joining:

• Surface heating (similar to heated tool/hot plate welding)
• Through Transmission Infrared Welding (TTIr)
• IR staking

These techniques are detailed in the following sections.

12.2.1   Surface Heating

This technique is similar to heated tool/hot plate welding. The surfaces of the components to be joined are heated by direct IR/laser exposure for a sufficient length of time to produce a molten layer, usually for 2–10 s. Once the surface is fully melted, the IR/laser tool is withdrawn from between the parts, the parts are forged together, and the melt is allowed to solidify.

The basic steps of surface heating with IR/laser welding are depicted in Fig. 12.1.

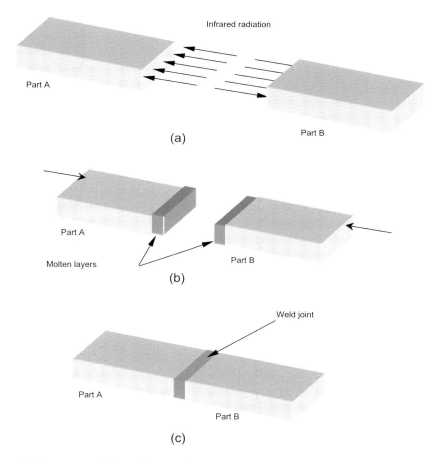

Figure 12.1 Basic steps of the IR/laser surface heating mode of welding (Courtesy of ETS, Inc.)

The six major steps of the IR/laser surface heating mode of welding are detailed below:

1. *Loading Parts.* This usually involves manually placing the parts into the fixture. In some cases, the fixtures are equipped with a vacuum to assist with securing the parts.

2. *Press Actuation*. This step is initiated by the operator and is usually accomplished by the closing of two palm buttons. The IR/laser source is translated between the two parts to be welded.

3. *IR Application*. Once the parts and the IR source are properly located, the IR radiation is applied to the parts. This promotes heating and melting of the faying surface (surface to be welded). The duration of the heating time is usually pre-selected by the operator.

4. *Change-Over Phase*. After the IR energy is applied to the parts, the laser source is translated out of the way and the parts are brought into intimate contact. The time between the radiation being discontinued and the parts being pressed together is often called the dwell time and must be minimized to assure that the outer layer of molten surfaces do not prematurely solidify.

5. *Clamp Phase/Hold Phase*. During this phase, the parts are held together under a preset pressure to assure intimate contact between the molten surfaces and to squeeze any contamination from the weld zone. Often mechanical stops are employed to restrict the movement of the parts and to prevent excessive melt displacement. Without the use of mechanical stops it is possible to squeeze nearly all of the melt out of the weld zone, resulting in a poor weld, often referred to as a "cold-joint".

6. *Unloading*. Once the molten material is sufficiently solidified, the parts are removed.

Table 12.1 provides typical cycle times required for laser surface heating welding.

***Table 12.1** Typical Surface Heating IR Welding Cycle*

Step	Typical time [s]
Loading parts	3 to 15
Press actuation	1 to 2
IR radiation application	1 to 5
Change-over time	1 to 2
Hold phase	1 to 3
Part unloading	5 to 10

12.2.2   Through Transmission IR Welding

This IR welding is based on the concept of passing IR/laser radiation (typically with wavelengths (λ) between 800 to 1050 nm) through one the components to be welded and having the second component absorb the energy at the interface (see Fig. 12.2). This absorption results in heating and melting of the interfaces and allows the parts to be welded.

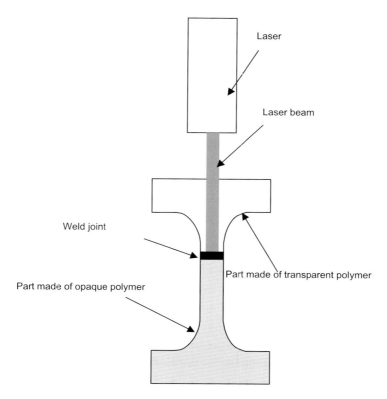

Figure 12.2 TTIr welding of plastics

The TTIr mode of IR welding is currently the most popular mode of operation because it offers several additional benefits compared to surface heating. For example, it is a pre-assembled method. This means, the parts are placed into the machine in the same position and orientation as the final, assembled position. For many applications, this is critical to allow sub-components to be held in place during the welding process without complex fixtures.

Other benefits of the TTIr mode include speed and flexibility. Typical cycle times range between 3 and 5 s. In comparison to hot plate welding, which has a typical cycle time of 10 to 30 s, thus TTIr is much faster. With the TTIr mode even unsupported internal walls with complex curvature can be welded, if the optical properties of the material allow illumination of the faying surface by IR/laser radiation. Applications with this type of geometry can be difficult to weld with vibration welding, a process that is considered relatively fast with cycle times less than 10 s.

Possibly one of the most important advantages of the TTIr process is weld quality. Because the process is non-invasive, the parts typically have excellent cosmetic properties. In addition, there are no excitation vibrations or large heated platens and only the weld area is heated and modified/melted.

One limitation to TTIr is material suitability. One of the components must be relatively transparent to the IR radiation. Since most TTIr systems on the market use a wavelength between 800 to 1,050 nm, most unfilled polymers tend to be transparent, however, crystalline polymers such as PE and PP tend to promote internal scattering of the radiation. This often limits the clear part to a thickness of less than 3 to 5 mm with scattering materials. More detailed information on this effect can be found in Section 12.6.

In addition to requiring that one part is transparent to the radiation, the other part must be absorbent. Usually this is accomplished by the addition of carbon black or another IR absorbing die. More information on material requirements is provided in Section 12.3.2 and in Section 12.6.

The five major steps in the IR/laser through transmission mode of welding are detailed below:

1. *Loading Parts.*   This usually involves manually placing the parts into the fixture. Often there is only one fixture, the lower fixture, which holds both parts.

2. *Press Actuation.*   This step is initiated by the operator and is usually accomplished by the closing of two palm buttons. The IR source, often referred to as the welding head, is translated to apply force and radiation to the parts.

3. *IR Application.*   Once the parts and the IR source are properly located, the IR radiation is applied to the parts. This promotes heating and melting of the faying surface (surface to be welded). The duration of the heating time is usually pre-selected by the operator. In some cases, the amount of melt or the displacement can be pre-selected.

4. *Clamp Phase/Hold Phase.*   During this phase, the parts are held together under a preset pressure to assure intimate contact between the molten surfaces.

5. *Unloading.* Once the molten material is sufficiently solidified, the parts are removed.

Table 12.2 provides typical cycle times required for through transmission IR welding.

Table 12.2 Typical TTIr Welding Cycle

Step	Typical time [s]
Loading parts	3 to 10
Press actuation	1 to 2
IR radiation application	1 to 10
Hold phase	1 to 3
Part unloading	5 to 10

12.2.3   IR/Laser Staking

IR staking is often used to join two materials, which cannot be welded, e.g., joining a metal component to a thermoplastic component. There are similar techniques, such as ultrasonic and hot air cold staking, however, IR staking is often selected because of its benefits, one of which is the fact that IR staking produces a small heat effected zone and does not produce any part marking.

Figure 12.3 shows a typical cross-section of a staking profile made with IR welding/joining. In the figure, the non-weldable material can be a metal, wood, or even a thermoplastic that is not compatible with the lower thermoplastic component.

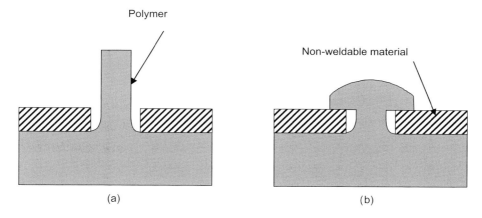

Figure 12.3 Typical cross-section of laser staking profile (Courtesy ETS, Inc.)

Figure 12.4 shows three studs that were staked with IR. In this case, a single head with three tools re-formed each stud simultaneously.

Figure 12.4 Typical staking application (Courtesy Branson Ultrasonics Corp.)

IR staking works in one of two modes: a single movement mode or a dual movement mode. In the case of the single movement mode, the IR radiation is transmitted through a special glass lens that acts as the top of the cavity, which defines the final shape of the stake. Figure 12.5 shows the typical stages for this mode of operation. The forming tool and the IR source are designed into a single component, thus allowing the process to occur in a single movement. This reduces the complexity of the process and typically makes it relatively fast. In addition, the molten material is squeezed out by the lens, exposing new un-melted material of the stud. This further decreases the cycle time since both mass flow and heat conduction promote heating of the stud.

The single movement mode of IR staking also allows a distance or collapse mode of control, resulting in a high level of repeatability from cycle to cycle. The controller can measure the initial height or position of the stud once the tool is in contact with the parts and reference this as a home or zero position. The controller then can continue to melt or displace the stud until a preset amount of displacement/collapse occurs.

However, since the lens, typically manufactured from glass or quartz, is in contact with flowing molten plastic, filler materials, such as glass fiber or metal, promote excessive tool wear. Therefore, this mode is usually not recommended for filled plastics.

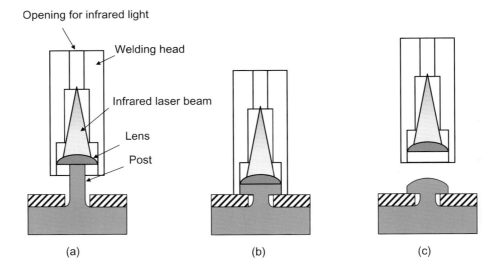

Figure 12.5 *Separate stages of single movement IR staking (Courtesy ETS, Inc.)*

If filled materials are used with IR staking, it is usually recommended that a dual movement process be used. Figure 12.6 shows the typical stages for this mode of operation where the forming tool and the IR source are designed into separate components, requiring two movements during the process. This allows the forming tool to be manufactured from a wide range of materials, including hardened steel. The dual movement process typically increases cycle time along with the fact that heat conduction is the sole heat transfer mechanism.

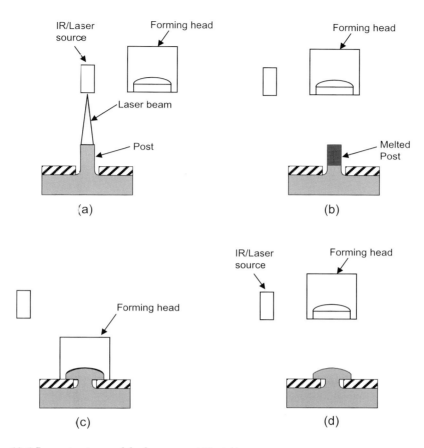

Figure 12.6 *Separate stages of dual movement IR staking*

Table 12.3 provides typical times required for the steps/stages for both, single and dual movement IR staking.

Table 12.3 *Typical IR Staking Cycle Times*

Step	Single Movement Process [s]	Dual Movement Process [s]
Loading parts	3 to 10	3 to 10
Press actuation	1 to 2	1 to 2
IR radiation application	3 to 15	3 to 15

Table 12.3 Typical IR Staking Cycle Times (Continuation)

Step	Single Movement Process [s]	Dual Movement Process [s]
Application of forming tools	NA (0)	1 to 3
Hold phase	1 to 3	1 to 3
Part unloading	5 to 10	5 to 10

12.2.4   Heating Configurations

When using IR welding, there are three primary methods to applying the IR radiation to the faying surfaces: scanning, continuous illumination and mask welding.

Scanning involves translating an IR source across the faying surface. This approach usually involves locating the parts in a fixture and translating the IR source with a robotic arm or a similar type of automaton. One of the main advantages of this approach is that a single welding machine can be easily re-programmed to weld a variety of applications. However, one limitation of this approach is the fact that the faying surfaces must be in intimate contact, because as the IR source translates around the circumference of the part, the faying surface is only locally heated, see Fig. 12.7. Thus, unless the application is compliant and can be locally deformed to force the faying surfaces together, any gaps between the faying surfaces can result in weld voids. Gaps between the two parts can be the result of mold warpage, mold shrinkage, ejector pin indication, or variations in the cavities.

In a scan welding technique, the typical diameter of an independent beam ranges from 0.6 mm to 2.6 mm, which allows the welding of intricate patterns as shown in Fig. 12.8.

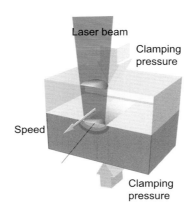

Figure 12.7 IR scanning and localized heating (Courtesy Leister Corp.)

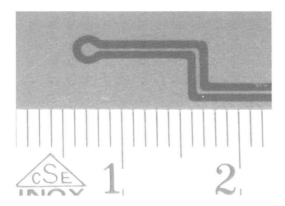

Figure 12.8 *Typical weld with scanning (Courtesy Leister Corp.)*

A basic square, circle, rectangle, or other contour can be sealed with a weld time of 2 to 5 s. Multiple laser sources can be used concurrently, allowing for batch processing and higher production speeds.

When welding a square pattern, or any pattern with sharp turns or bends, the temperature of the plastic in the corner or sharp contour area will be higher. Precise Proportional Integral Differential (P.I.D.) control can be utilized with the single beam method to control or reduce the intensity of the light as the beam passes over the intricate corners or contours. This will eliminate over-welding areas and ensure an even weld width throughout the weld pattern.

In many instances, especially with microfluidic or sensing devices, displacement or collapse of the substrates is not tolerable, as it will affect the fluidic channel size or final part height. Inherent to the scanning technique the benefit of a fixed or mechanical stop produced by the un-melted regions of the parts, providing identical and consistent stack heights before and after the welding process. Thus, scan welding is well suited where collapse or displacement is not desired.

In scan welding with the TTIr mode, small voids can be welded or sealed because of thermal expansion as the plastic heats and melts. The size of the void that can be fully welded or sealed depends on the welding parameter as well as on material properties, such as the coefficient of thermal expansion (CTE). Figure 12.9 shows weld strength as a function of defect size for welds made with ANSI (AWS G1-Standard) samples. It is seen that there is a significant loss in weld strength with defects as small as 0.25 mm [1]. The material in this study (aliphatic Polyketone) is a high-crystalline material with a relatively high CTE, (25 to 55 °C) 1.9×10^{-5} cm/cm/°C.

When using IR scanning, the light source can be translated around the part at a rate so that the beam returns to any given point before the material is allowed to solidify. This allows the entire faying surface to be melted during the weld cycle. Unless the part is relatively small, the minimum scanning rate requires the beam to be translated with galvanic mirrors or a very fast robotic system. If galvanic mirrors are utilized, the IR source is usually limited to a laser source in order to keep the beam collimated as it reflects off the mirrors. This mode has the advantage that melt-down or collapse is possible. Thus, virtually any defect size can be welded or sealed with the proper amount of melt-down. It also retains the benefit that a single welding machine can be easily re-programmed to weld a variety of applications. If galvanic mirrors are used however, the application must be relatively planner and no "shadow" areas can be tolerated. That means that the part geometry must allow the entire faying surface to be illuminated from a single point, using the rotating mirrors.

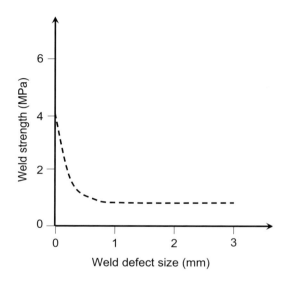

Figure 12.9 *Example of weld strength as a function of weld defect for IR welding by scanning (Courtesy Branson Ultrasonics Corp.)*

The second possible heating mode for TTIr is continuous illumination. In this mode, there are several IR sources illuminating the entire faying surface during the weld cycle, see Fig. 12.10. As with high speed scanning, part tolerance and fit-up are not critical because melt-down is possible. In addition, complex geometries can be welded without the limitation of shadow areas and there are no problems with "run-on/run-off"

Another possible heating mode for TTIr, is mask welding. This technique utilizes a typical continuous illumination curtain of light that is passed over a mask blocking or masking portions of the light, allowing only the pre-specified areas to melt and seal. An example of this method is shown in Fig. 12.11. This method is especially suited for complex, micro-structured areas. Microwelds as narrow as 100 μm can be achieved (Fig. 12.12). The correlation of light intensity, clamping force, and travel speed (the speed of the laser passing over the mask or the mask assembly passing under the laser) controls the amount of melt and edge definition. Figure 12.13 graphically illustrates the correlation between travel speed and width of the weld line.

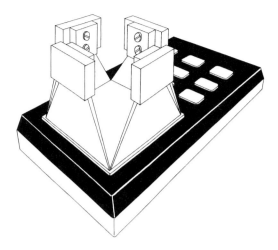

Figure 12.10 *Continuous illumination with IR welding (Courtesy Leister Corp.)*

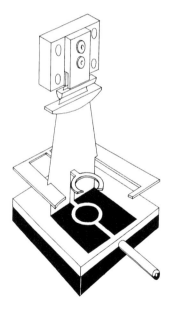

Figure 12.11 *Example of TTIr mask welding (Courtesy Leister Corp.)*

As with TTIr, when using surface heating, there are also various methods of introducing heating: scanning (high speed) and continuous [2]. The heating source for continuous illumination with surface heating is very similar to the one previously described for TTIr. However, since surface heating relies on residual heat and melt at the faying surfaces, slow speed scanning is not applicable. Instead, only high speed scanning can be used to build up a sufficient melt layer. In this case, the beam is often split with a mirror to illuminate both parts simultaneously, as seen in Fig. 12.14. The rotating mirror usually dithers back and forth to direct the beam from one secondary mirror to the other. In addition, it is possible to rotate the secondary beam to increase the width of the heated area.

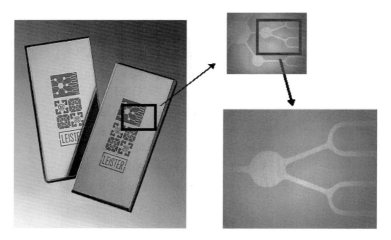

Figure 12.12 Example of microwelds with TTIr mask welding (Courtesy Leister Corp.)

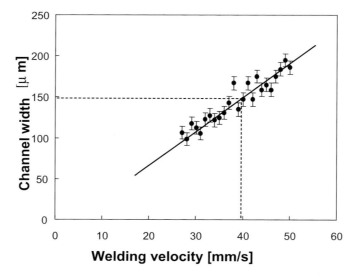

Figure 12.13 Example of possible weld width as a function of travel speed (TTIr-mask welding) (Courtesy Leister Corp.)

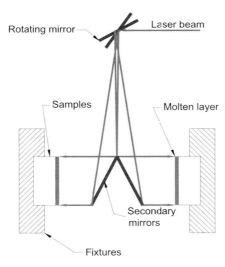

Figure 12.14 Surface heating via scanning

12.2.5   IR Welding and Cutting

In some applications, primarily films and fabrics, IR welding has not only been utilized to weld and seal two parts together, but to also cut along the edge (or toe) of the weld. This has the advantage of increased productivity, because two steps (welding and cutting) are combined into a single phase. In addition, the resulting cut edges are usually cauterized and relatively smooth compared to those achieved by mechanical cutting with shears, see Fig. 12.15.

One limitation of IR welding and cutting is the generation of fumes. Usually the machine design must incorporate a filtration system to extract any air-born particles or fumes, because in some cases, the resulting fumes and particles can be hazardous. The reader is directed to the Center for Devices and Radiological Health (CDRH) [3] for additional information on air-born hazards with laser cutting.

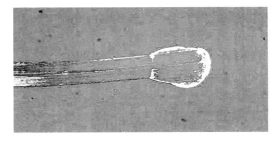

Figure 12.15 Cross section of film application, cut and sealed with laser (Courtesy TWI, Cambridge, UK)

12.3   Physics of the Process

12.3.1   The Electromagnetic Spectrum

While the human eye is only sensitive to wavelengths between 400 nm (violet) and 780 nm (red), the electromagnetic spectrum is much broader, see Figure 12.16.

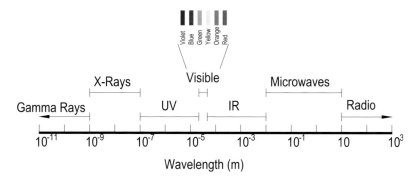

Figure 12.16 Electromagnetic spectrum

In order to understand IR and laser welding of plastics it is important to understand the importance of the electromagnetic spectrum, in particular of those wavelengths (λ) between 800 and 11,000 nm. Wavelengths between 800 and 11,000 nm are the most common for laser/IR plastic welding and processing. As will be detailed in the following sections, there are several material characteristics, which make these wavelengths more amendable for processing. In addition, these wavelengths are also easily produced at relatively high power levels, >10W. For example, one of the most common lasers used in industry, a CO_2 laser, produces radiation at a wavelength of 10,600 nm. Another common high power laser, YAG (yttrium-aluminum garnet) laser, produces a wavelength of 1,064 nm [4]. In the late 1990s, high power laser diodes became common. Laser diodes typically produce wavelengths between 800 and 950 nm. In addition, non-laser sources such as ceramic heaters and quartz lamps are able to produce similar wavelengths, 5,000 to 10,000 nm and 1,000 to 5,000 nm respectively.

Shorter wavelengths can be produced, but only at relatively lower power levels. For example, a typical excimer laser can produce wavelengths as short as 200 nm, however, the power levels are usually less than 10 W for a single laser. Because of the relatively low power capacities, these wavelengths are usually limited to very small components or surface modifications in terms of plastic processing.

One of the reasons why it is more difficult to produce the shorter wavelengths is that the amount of energy in the electromagnetic spectrum is inversely proportional to the wavelength, see Eq. (12.1). Here, h is Planck's Constant (6.626×10^{-27} erg s) and c is the speed of light (3.0×10^{10} cm/s) [5].

$$E = \frac{hc}{\lambda} \tag{12.1}$$

Thus, as the wavelength (λ) decreases the energy (E) increases and it becomes more difficult to produce the radiation.

12.3.2   Molecular Absorption and Composition Interactions

There are three basic interactions between electromagnetic radiation (EMR) and polymers: reflection (R), absorption (A), and transmission (T). It is beyond the scope of this book to review each of these interactions in detail, but there are several key aspects, that are important to plastic welding. For example, it is common to define the amount of each interaction as defined in Eq. (12.2).

$$R = 100\% \frac{I_R}{I_O}$$

$$T = 100\% \frac{I_T}{I_O} \tag{12.2}$$

$$A = 100\% \frac{I_A}{I_O}$$

where I_o is the intensity of the radiation on the plastic directed perpendicular to the surface. I_R is the intensity of the light, which is reflected by the plastic. I_T and I_A are the intensities of the light, which passes through and is absorbed by the plastic, respectively.

While the amount of each possible interaction varies greatly for each plastic and wavelength, the total of all three must be unity as defined in Eq. (12.3).

$$100\% = R + T + A \tag{12.3}$$

In terms of plastic joining, absorption and transmission typically have the greatest effect on the process. While reflection can have some effect with plastics, it is usually limited to coated applications, such as aluminized applications used in lighting applications. Most homogeneous plastics do not reflect a significant amount of EMR (light in this example), usually between 2 and 8%. This is caused by the fact that most plastics have an index of refraction between 1.4 and 1.6 [6]. The fraction of light being reflected (R) is defined in Eq. 12.4. Here, n is the index of refraction of the plastics and m is the index of refraction of air ($\sim$1).

$$R = \frac{(n-m)^2}{(n+m)^2} \tag{12.4}$$

Determining the amount of transmission is more complex. As light travels from one material (or medium) to another, it can follow a straight path only if the path is perpendicular to the interfaces or if the relative index of refraction of the interfaces is equal. If the light ray is not perpendicular to the interface, as in Fig. 12.17, the path is refracted.

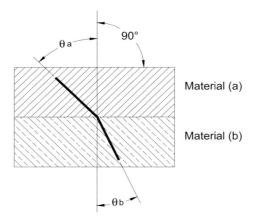

Figure 12.17 *Transmission of light through two materials with different optical densities*

The amount of refraction is defined by Snell's law, Eq.12.5, where n_a and n_b is the index of refraction for material a and b respectively. The effect of refraction is important to under-stand in applications where it is not possible to direct the radiation directly at the weld area or where the beam/wave must pass through multiple layers of plastic to reach the weld area. This type of interaction is typical when working with amorphous materials such as PC, PS, and Acrylics (optically clear plastics).

$$\frac{\sin \theta_a}{\sin \theta_b} = \frac{n_a}{n_b} \qquad\qquad (12.5)$$

In crystalline materials the light generally follows a similar path as previously described [7], however the structure of materials, such as PA, PE, and PP, causes internal refraction and scattering, see Fig. 12.18. Each phase, crystalline and amorphous, has a different index of refraction and thus light is refracted as it travels through the sample. In the case of a crystalline thermoplastic, or more accurately a semi-crystalline thermoplastic, a beam will encounter a nearly endless number of phase changes within a sample.

This internal scattering only affects a material's weldability when TTIr is used. The scattering causes the EMR to diffuse as it travels through the sample, reducing the effective energy that reaches the faying surface. The amount of scatter is defined by Lambert–Bouger's Law (Eq. 12.6), in which I_t is the intensity of the light at a thickness, t, (or depth) through a sample with an absorption constant of α. If it is assumed that scatter is can be similarly described it is seen that thicker samples produce more internal scatter. Thus, the energy required to melt the interface, is proportional to the thickness of the sample when welding diffusive materials, such as crystalline thermoplastics [8].

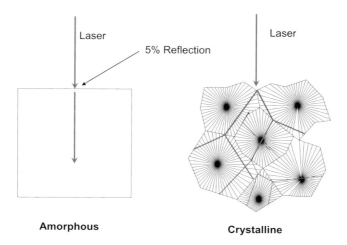

Figure 12.18 *Light transmision through amorphous and crystalline plastics*

$$I_t = I_0 e^{-(\alpha t)} \qquad\qquad (12.6)$$

The amount of light that is absorbed by a pure, unfilled, thermoplastic is defined by the material's chemical structure. For example, molecular bonds can be excited to vibrate in particular modes by absorption of IR radiation. Many bonds can absorb different wavelengths and bend in different modes. While there are many different possible modes of vibration, two commons modes are stretching and bending, see Fig.12.19.

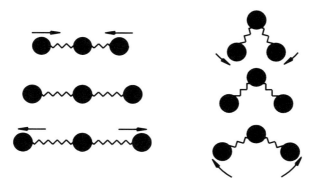

Figure 12.19 *Possible modes of bond vibrations: (a) stretching (b) bending*

Historically, the relationship between absorption and chemical structure has been used for material identification. By exposing a sample to a wide bandwidth IR radiation (typically a wavelength between 3,000 to 15,000 nm) and observing the absorption, it is possible to determine which chemical bonds are present [9], see Table 12.4, and to interpret a material's chemical structure.

Table 12.4 *Selected Modes of Vibration for Some Chemical Bonds and IR Absorption [5]*

Bond	Group	Mode	λ [μm]
C-H	CH_2, CH_3	Stretching	3.3–3.4
C-H	$\equiv$C-H	Stretching	3.0
C-H	-CH_2-	Bending	6.8
O-C	>C=O	Stretching	5.4–5.9
O-H	-O-H	Stretching	2.7–2.8

While the relationship between absorption and molecular structure is advantageous for chemical analysis, it makes most plastic materials non-transparent to IR wavelengths above 1,000 nm (1 μm), except for varying selected wavelengths. As seen in Fig. 12.20, common un-filled plastics are relatively transparent from 0.4 to 1.1 μm. Thus, when welding plastics with TTIr, wavelengths below 1.1 μm are preferred. In addition, the IR radiation can be easily generated with laser diodes, YAG laser, and quartz halogen lamps. It is important to remember that most unfilled plastics are relatively transparent between 0.4 to 1.1 μm, but as previously discussed, crystalline material promote internal scatter.

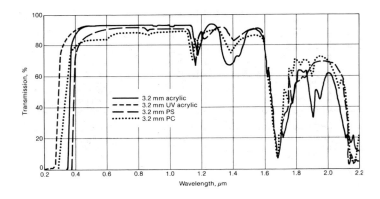

Figure 12.20 *Transmission of common optical plastics (Courtesy AMS International [10])*

It is also important to keep in mind that scattering is promoted not only by the physical structure of a plastic, but that additives have a similar effect. Additives, such as glass fibers, talc, and in-organic dies and pigments can promote internal scatter. Organic dies tend to dissolve within plastics and usually have a particle size (or molecule) much smaller than the wavelength of light (EMR), and therefore do not promote refraction [11]. That is to say, shorter wavelengths or larger particle sizes tend to promote more scatter.

In order to make plastic absorb IR radiation, particularly wavelengths below 1.0 μm, additives are used in the base material. The most common additive is carbon black, which effectively acts as a black body and absorbs nearly all wavelengths. The concentration of carbon black required to make a plastic absorb IR radiation is usually less than 1%.

The amount of carbon black greatly affects the depth heating [12]. Amounts of carbon black as low as 0.03% can result in near 100% absorption of EMR within a depth of 0.5 mm of the plastic. Alternatively, 0.07% carbon black can absorb 100% of EMR within a depth of 0.25 mm.

The depth of heating is generally inversely proportional to the amount of carbon. If too much carbon black is added to a plastic, it is possible to limit the heating exclusively to the surface. As seen in Fig. 12.21, if a relatively high amount of carbon black is added to a plastic, the surface reaches a relatively high temperature with little heating below the surface. In contrast, if a relatively low amount of carbon black is used, the temperature distribution is wider. In this case, the heated layer reaches deeper into the sample and the peak temperature is lower.

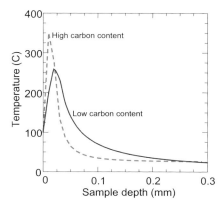

Figure 12.21 Hypothetical temperature profile for two relative levels of carbon black

Surface heating caused by excessive carbon black can promote several undesired side effects. If all the energy is absorbed at the surface, the energy density can be so high that the peak temperature can promote polymer degradation. Porosity is another unwanted side effect for hygroscopic plastics. It can occur because the vapor pressure of moisture in solution in a plastic is proportional to the samples temperature [13]. Thus, as the bondline temperature increases above the melt temperature of the plastic, the vapor pressure of the water increases for a given moisture content. If the bondline pressure (clamp pressure) is below the resulting vapor pressure, porosity will form, if sufficient time for molecular diffusion is allowed. The time required for this diffusion depends on the material, temperature, and morphology, but is usually relatively short, less than 5 s [14]. The generated porosity can act as stress points and greatly reduce weld strength. Excessive porosity can also provide leak paths for applications that require the weld to be hermetic. In most cases, porosity is usually detrimental to the weld performance.

Higher temperatures can also promote excessive squeeze flow of the melt because viscosity is inversely proportional to temperature. Excessive squeeze flow can produce undesired flash and promote molecular alignment that is perpendicular to the weld joint. This type of molecular alignment can produce relatively high residual stresses and reduces the weld strength.

As previously discussed, relatively low concentrations of carbon black can promote deeper and more distributed heating. This type of temperature distribution usually results in a more moderate peak temperature and reduces the potential of polymer degradation, formation of porosity, and excessive squeeze flow. In addition, deep heating promotes a thicker bondline, which in turn promotes better molecular alignment and entanglement.

12.3.3   Heating Equation

If it is assumed that all of the energy transmitted from a laser source is absorbed at the faying surface, that is A=100%, it is possible to estimate the amount of power, or heat flux, required to melt the interface. Assuming that heat losses at the surface due to radiation and convection are negligible, it is possible to estimate the time to melt the surface and the temperature distribution within a weld for a given heat flux (q_o) [15]. If a continuous illumination mode for laser/IR welding is used, it is possible to estimate the temperature, θ, at any location and any time by using Eq. 12.7 [16].

$$\theta(x,t) = \theta_i + \frac{2 \cdot \dot{q}_0}{\lambda} \left[\sqrt{\frac{\kappa \cdot t}{\pi}} \cdot \exp\left(-\frac{x^2}{4 \cdot \kappa \cdot t}\right) - \frac{x}{2} \cdot \operatorname{erfc}\left(\frac{x}{2\sqrt{\kappa \cdot t}}\right) \right] \tag{12.7}$$

Where θ is the temperature, x is the position, t is time, θ_i is the initial temperature of the solid, λ is the thermal conductivity, κ is the thermal diffusivity, and erfc(z) is the complementary error function. It is important to note that this model is good for predicting temperature distributions within a semi-infinite body where there are no temperature gradients in the thickness (z) direction. This model is further detailed in Chapter 2.

If a scanning technique is used, Eq. 12.7 cannot be utilized to predict temperature profiles. With a scanning mode, the heat source is applied to a small region of the faying surface. In addition, the heat source is moving. Thus an alternative model, the Rosenthal's model, must be used. Rosenthal's point [17] source theory analyzes the temperature distribution in a semi-infinite plate with a point heat source moving at a constant velocity (see Fig. 12.22).

Two coordinate systems are considered, a fixed coordinate system and a moving coordinate system with its origin located on the surface of the plate below the heat source. The moving coordinate system moves at the same velocity (v) as the heat source. Neglecting edge effects, this allows the model to be greatly simplified since the problem is reduced to a quasi-steady-state heat flow condition. For example, far away from the edges (start and end of weld), the temperature distribution in the moving coordinate system is constant at all times. A point located 1 cm behind the heat source and moving with the heat source, will always remain at the same temperature at all times. Relative to the fixed coordinate system, points experience a heating and cooling cycle as the heat source passes near any given point. However, since the moving coordinate system is moving with the heat source, points in the moving coordinate system do not experience heating and cooling. In the moving coordinate system, w is related to the x-coordinate and is defined as:

$$w(t) = x - vt, \text{ and } r(t) = \sqrt{w(t)^2 + y^2 + z^2} \tag{12.8}$$

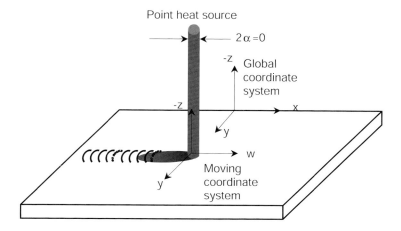

Figure 12.22 *Details of moving heat source with moving coordinate system*

By using the moving coordinate system, Rosenthal showed that it is possible to calculate the temperature at any location within the plate using the equation:

$$\theta(t) = \theta_i + \frac{P}{2\pi\lambda r(t)} \exp(\frac{-v}{2\kappa}(w(t) + r(t))) \tag{12.9}$$

where θ_i is the initial temperature of the body. It should be noted that in this model it is assumed that the laser or IR source is focused to an infinitesimally small spot so that it can be approximated by a point. That is to say the focal radius, α, is zero. If the heat is not well focused it cannot be modeled as a point heat source. In this case, a distributed heat source model like the one developed by Eagar and Tsai [18] must be used. In many cases the point heat source model can be used to predict temperature distributions within a plastic weld made with focused and moving IR/Laser source.

12.4   Equipment Description

12.4.1   Equipment Components

IR welding systems vary greatly in design and complexity depending on their mode of operation. However, there are several components common among most machines:

Generator/Power Supply: This component converts line voltage and frequency into the correct voltage, current, and frequency to power the IR source. As detailed in Section 12.4.2, there are many different IR sources and each has different power requirements. The selection of the source is usually based on the equipment manufacturer's preference. Each source has it own benefits and limitations, which are detailed in Section 12.4.2.

Control: The control establishes an interface between the machine and the operator and monitor the system. They usually contain a PLC or similar type of logic circuitry, interfacing the logic circuits with the user. The controller also reports to the operator on the condition of the machine, such as welding data and machine status. The controls will vary greatly depending on mode of operation and machine design. For example, if the machine scans the laser, the controller will have some type of learning/teaching routine to program the robotics. If the machine design illuminates the entire surface simultaneously, the controller will only allow the operator to vary such parameters as weld time, laser power output, and melt depth (displacement or collapse).

Actuator: This component is usually a pneumatic press that moves the upper fixture (platen) to the lower fixture and applies force during the weld and hold cycle. There is usually a pressure regulator, which determines the maximum force applied to the press. Some machine designs incorporate a servo-based actuator to generate the movement and force. Servo systems are more popular in machines, which illuminate the entire surface simultaneously, where control of displacement or collapse is possible. Often, the actuator will incorporate an encoder in order to measure and report the amount of collapse.

Lower Fixture: In many cases the lower fixture is a simple jig, which locates the parts being welded. Because IR/laser welding is often used for critical applications, which have a relatively low degree of tolerance, care is required to assure the fixture provides proper alignment and is properly located.

Upper Fixture: This component is often the most complex and most critical component of the entire machine. Typically, it is within this component that the IR/laser radiation is generated and delivered to the application being welded. The upper fixture varies greatly from machine to machine and is dependent on the machine design concept, applications, and heating configuration, see Section 12.2.4. For each heating configuration, there are several typical designs:

- With conventional scanning (or slow speed scanning), the laser is typically translated via a 2- to 5-axis robotic system.
- If the IR source is a CO_2 laser, often the part is translated, because focusing of the beam is critical.

- If the IR source is a YAG laser or laser diode, it is possible to use fiber optic cables which can be easily translated. In this case, the part remains stationary.

- If high speed scanning is used, the beam is usually translated over the faying surface with the help of galvanic mirrors. In some cases, it is possible to translate the beam in a high-speed mode by the use of robotics and servos, depending on the size and complexity of the application.

- If the machine operates by illuminating the entire surface simultaneously, there are several different possible head designs. It is possible to focus the array of IR sources directly at the faying surface, called direct coupling. In this case, the sources (IR lamps or laser diodes) are positioned above the application and follow the contour of the faying surface. An alternate approach is to couple the IR/laser source to the application through fiber optics. This allows the diodes to be remotely located, thus allowing for easier maintenance because the lasers are protected from dirt and contaminants. The major disadvantages of fiber coupling are costs and losses due to fiber packing and reflection.

IR/Laser Enclosure: In nearly all equipment designs the operator needs to be protected from the radiation by some type of enclosure. If the machine contains laser(s), the enclosure needs to be FDA certified in order to assure that the operator is properly protected. Lamps, such as quartz halogen, generally produce a much lower power density and have a high degree of divergence, which inherently makes them slightly safer compared to lasers. however, the operator needs to be shielded from the intense light.

Figure 12.23 shows a typical IR/laser machine, which works by illuminating the entire surface simultaneously. It is important to note that there are no port holes for viewing the operation due to safety reasons. However, in this particular model, there is a monitoring system that allows the operator to watch the process via a special camera that is sensitive to the IR radiation.

Figure 12.23 *Typical IR/laser machine for simultaneous illumina-*
tion (Courtesy Branson Ultrasonics Corp.)

Figure 12.24 shows a system used for the scanning technique. The overall machine size is usually larger compared to the one seen in Fig. 12.23, in order to accommodate the robotics and/or mirror systems. Again, a vision system is incorporated into the design that allows the operator to view the welding operation. Figure 12.25 shows a typical system for TTIr mask welding.

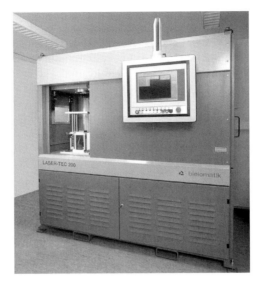

Figure 12.24 *IR/laser system for scanning mode (Courtesy Bielomatik)*

Figure 12.25 *Photograph of TTIr-mask welding system (Novolas ™*
μ-Microsystem, Courtesy Leister Corp.)

12.4.2   IR Sources

As detailed in Section 12.3.1, the electromagnetic spectrum has a broad range of wave-lengths. It is common to define the spectrum into various regions. For example, light with wavelengths between 2,000 to 30,000 nm is referred to IR and near-IR range between 760–2,000 nm. Each wavelengths can be generated by a number of different sources. In fact, a wavelength of 10,600 nm can be generated by a CO_2 laser or by a ceramic heater. However, there are several fundamental differences between the different types of radiation generated. For example, CO_2 laser radiation is usually relatively collimated (beam shape), coherent, and monochromic (narrow wavelength distribution). In contrast, the radiation generated by a ceramic heater at approximately 270 °C radiates in all directions, is not coherent, and has a relatively wide wavelength distribution (poly-chromatic), see Fig. 12.26. In some applications, the broad output band-width of non-laser sources can produce unwanted side effects, such as undesired heating. In these cases, a filter must be employed to absorb the undesired wavelengths. These filters can be standard glass filters, however, because of the heat buildup within the filter, heat removal is usually critical. Cooling can be obtained through a flow of air or water. In some critical applications, where filtering must be perfectly matched to the application/material, a liquid filter [19] can be used. In this case, a liquid is selected, which has very similar optical properties compared to the plastic, such as a pre-polymer or monomer, and this liquid is passed through a sealed chamber between the lamps and the application. This type of filter also has the advantage that it is possible to pass the liquid directly through a heat exchanger and recycle it contin-uously.

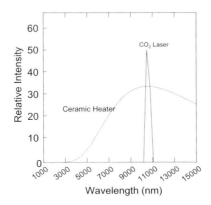

Figure 12.26 *Hypothetical comparison of radiation as a function of wavelength for CO_2 laser and ceramic heaters*

It is also possible to generate shorter wavelengths with either lasers or lamps, e.g., it is possible to generate 800 to 900 nm with laser diodes and 1064 nm with a YAG laser. Both sources produce relatively well-collimated beams depending on the optics/lens incorpo-rated. Both sources also produce monochromatic radiation. It is important to note however, that laser diodes generally do not produce coherent radiation. This difference usually has very little impact on most plastic welding applications. Similar wavelengths can be generated with quartz halogen lamps, which are offered in standard packages with different reflectors, however one of the more common and useful is the MR-16 [20]. With this configuration it is possible to cut the reflectors edges to produce a relatively straight line of radiation, see Fig. 12.27. Usually standard lamps must be coated with a highly reflective material, such as aluminum in order to reflect the proper wavelength. Most bulb manufactures coat the reflecting surface with a special material, which allows wavelengths above 800 nm to be transmitted out the back of the bulb's reflecting surface to reduce heating of the illuminated target. However, this design is not desired in welding applications.

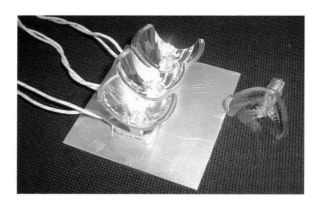

(a)

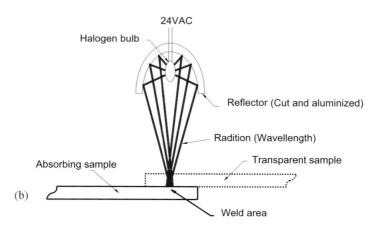

(b)

Figure 12.27 *(a) MR-16 Quartz halogen lamps with edges cut for tight packing and (b) cross sectional view of ray tracing of radiation and possible focusing (Courtesy Branson Ultrasonics Corp.)*

Compared to ceramic heaters, the MR-16 package allows for relatively good focusing of the beam, as seen in Fig. 12.27(b). Although the focusing is not as well defined with lasers, it is often sufficient. The radiation from quartz halogen lamps is also poly-chromatic and filtering may be necessary depending on the application.

Among the numerous types of commercially available lasers, CO_2, YAG and laser diodes are the most commonly used for plastic processing and welding. There are also countless IR lamps/heaters available but the most commonly used ones are quartz halogen, quartz, and ceramic lamps. Each of these sources has unique characteristics that usually determine which source is best suited for a particular machine design and application. Some of these characteristics are detailed in Table 12.5

Table 12.5 Typical Characteristics of Common IR Sources

IR Source	Wave length [nm]	Typical power range [W]	Divergence	Costs [$ per W]	Applications	Typical head designs
Laser diode	800 to 950	1–60 per device	1 to 10°	~50	Marking, heating and welding of opaque plastics	TTIr, Fiber coupled heads
YAG laser	1064	10–100	Less than 1°	~500	Marking, heating and welding of opaque plastics	TTIr, Fiber coupled and scanning heads
CO_2 laser	10,600	100–10,000	Less than 1°	~50–100	Marking, heating and welding of most plastics	Scanning heads
Quartz-halogen lamp	Peak: 1000–1500	50–300 per device	20–90°	~10	Heating and welding of opaque materials	TTIr, scanning heads
Quartz-lamp	Peak: 1500–5000	50–2000 per device	20 to 180°	~5	Heating and welding of most plastics	Area and line heating heads
Ceramic heater	Peak: 5000–10000	50–2000 per device	~180°	~5	Heating and welding of opaque materials	Area and line heating heads

12.5   Applications

There are many advantages to IR/laser welding including:

- Speed (3–10 s/weld)
- Relatively low residual stresses in welded parts
- No part marking
- No particulate generation
- No vibrations or movement that can promote damage
- Able to weld to large collapses in the continuous illumination mode
- Easily automated

When IR/laser welding is the selected method for joining of an application, usually one of the above benefits provides justification for the capital costs for the equipment relative to other joining techniques, such as ultrasonic welding and heated tool welding.

Some of the initial applications considered for welding with IR/Laser were parts such as instrument panel clusters, head lights, and marker lights for the automotive industry, because these applications require a hermetic seal without any marking or particulate generation. Because the application had an optically clear component (lens) and a relatively opaque component (housing), it was well suited for the TTIr mode of welding, see Figure 12.28. The only major consideration when using TTIr for this type of application, is to make sure that, when the housing is aluminized (for reflectivity), the aluminum does not coat the faying surface and prevent proper heating during the welding operation.

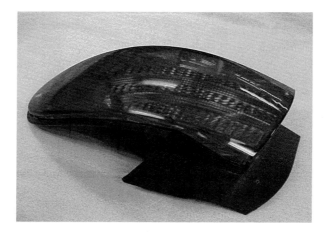

Figure 12.28 *Typical taillight application for TTIr mode IR/laser welding (Courtesy Branson Ultra-sonics Corp.)*

One of the main reason for considering IR/laser welding for head light and marker light applications is the reduction of residual stresses [21]. The residual stresses from welding, and sometimes even from molding, can lead to stress cracking when the part is exposed to solvents, such as gasoline or cleaning agents. The cracks can lead to leaks or even total separation of the two components. The amount of residual stresses induced by IR welding amounts to approximately 50% of the stress induced by ultrasonic welding, depending on the process parameters, see Fig. 12.29. This graph is only an example of typical stress generation for these welding processes and experiments must be conducted for each application and material combination for quantitative comparisons.

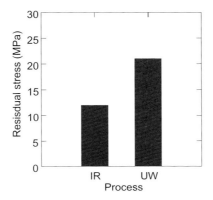

Figure 12.29 *Comparison of residual stresses induced by IR and ultrasonic welding (UW) (Courtesy Branson Ultrasonics Corp.)*

The fact that IR/laser welding is a pre-assembled joining method, i.e., the parts are placed into the machine in the same position and orientation as the final, assembled position, is another benefit of this welding method. For many applications, this is critical to allow sub-components to be held in place during the welding process without complex fixtures.

Other benefits of IR/laser welding include speed and flexibility. A typical cycle time ranges between 3 and 5 s, while hot plate welding has a typical cycle time of 10 to 30 s. With IR/laser even unsupported internal walls with complex curvature can be welded. Applications with this type of geometry can be difficult to weld with vibration welding, a process that is also considered relatively fast with cycle times of less than 10 s.

Possibly one of the most important advantages of the IR/laser welding process is weld quality. Because the process is non-invasive, the parts typically have excellent cosmetic properties. Since there is no relative motion between the parts, no excitation vibrations or large heated platens, only the weld area is heated and modified/melted. In short, "part marking" or "bleed-through" can be prevented. In addition, it is possible to accurately control the power dissipation within the weld, because the process can be easily controlled by varying the power of the IR/laser radiation. This allows control of the melt and welding process. The end result is less flash and no particulates.

Case Study 1: CO Detector Filter

This application had a thin filter medium attached to the housing and when alternative techniques were used, such as ultrasonic welding, this filter medium was often destroyed. However, with IR welding the application was welded without any damage to the filter and generation of particulate, see Fig. 12.30.

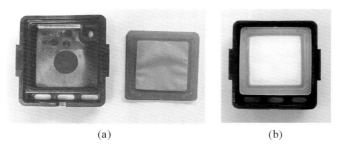

(a)   (b)

Figure 12.30 CO filter tester welded with IR/laser to prevent part damage, (a) un-welded, and (b) welded (Courtesy Quantum Group Inc.)

Case Study 2: Film and Fabric Applications

The photograph in Fig. 12.31 shows a medical application. In this IV-bag, blood is exposed to the weld, which makes it critical for patient safety to avoid particulate generation. Laser welding allows such complex geometries to be welded without any particulate generation. In addition, with proper equipment setup, it is also possible to weld applications of this type without flash generation. In medical applications with blood contact, flash generation can cause turbulent flow of the blood and result in damage to the blood platelets.

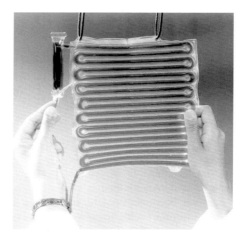

Figure 12.31 Film application with IR/laser welding (Courtesy TWI)

Case Study 3: High-Speed Cut and Seal Film (300 m/min)

Very high-speed cut and seal film (300 m/min) applications have been completed with CO_2 lasers. The edges of the cut seam are usually cauterized and show either no or only little fraying of the edges [22] .

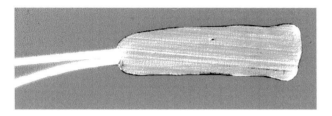

Figure 12.32 Cross section of material cut and sealed at high speed (Courtesy TWI)

Case Study 4: Break Fluid Reservoirs in Automotive Applications

Break fluid reservoirs in the automotive industry are another typical application where particulate generation cannot be tolerated. With the introduction of anti-locking break systems, valve and channel size for fluid transmission became much smaller compared to previous designs. This made minimum particulate generation critical to prevent clogging of the fluid channels. Figure 12.33 shows a photograph of such an application. It is important to note that a white lid is shown in this photograph for purposes of representation. However, another configuration is where both components appear to be black even though this application was welded in the TTIr mode. Here, a special die was added to the lid of the reservoir, which is IR transparent but appears black in the visible spectrum. Additional information on dies is provided in Section 12.6.

Figure 12.33 Break fluid reservoir welded with TTIr

There are some limitations restricting the use of IR/laser welding in some applications, such as:

- Material compatibility
- Application thickness for TTIr welding mode with crystalline materials
- Application optics must allow illumination of the faying surface (no shadow lines)
- Capital costs tend to be higher compared to alternative joining techniques.

All of these must be fully considered.

12.6   Material Weldability

Since there are several modes of IR/laser welding, it is not possible to define weldable materials for the process in general. This definition can only be given for each particular mode. In addition, since IR/laser welding is just now becoming more popular in industry, information on material weldability is rather limited. Therefore, any material considered for IR/laser welding should be tested first.

Some of the materials successfully welded to date include:

- PC
- PMMA
- EVOH [23]
- Acrylic
- PS
- ABS [12]
- PVC
- PE
- PP
- PK [1]
- Elastomers
- PA [11]
- Acetal
- TPFA [24]

Table 12.4 shows the general compatibility of these materials with the different modes of IR/laser welding. It is important to remember that any material should be tested for compatibility with each mode before selecting IR/laser welding. In general, one component must be absorbent to IR radiation and unless relatively long wavelengths are used (>1,500 nm), an absorbing additive, such as carbon black must be used with one of the components. In addition, if TTIr is used, at least one of the components must be IR transparent. As a general guideline, if a material is transparent in the visible spectrum it is also likely to be transparent in the near-IR spectrum. For example, materials such as PC and PS tend to be nearly 100% visibly transparent and they are also nearly 100% transparent to IR produced by laser diodes and YAG lasers. In addition, crystalline materials such as PE and PP are relatively transparent to near-IR radiation; however, because of their scattering light characteristics it is difficult to pass sufficient radiation through samples thicker than 5 mm to make a weld. It is possible to weld thicker samples, but relatively high power densities are required.

Table 12.4 Typical Material Compatibility with IR/Laser Welding Techniques

Material (Unfilled)	TTIr	Surface Heating	Staking
PC	Good	Good	Good
PMMA	Good	Good	Good
EVOH	Good	Good	Good
Acrylic	Good	Good	Good
PS	Good	Good	Good
ABS	Transparent grade only	Good	Good
PVC	Transparent grade only	Good	Good
PE	Thickness < 5 mm	Good	Good
PP	Thickness < 5 mm	Good	Good
PK	Thickness < 5 mm	Fair	Fair
Elastomers	Thickness < 5 mm	Fair	Fair
PA	Thickness < 5 mm	Fair	Fair
Acetal	Thickness < 5 mm	Fair	Fair
PTFE	Thickness < 5 mm	Poor	Poor

12.6.1   Effects of Additives and Fillers

It is possible to increase the transmission of IR/laser radiation through a crystalline material by modifying the micro-structure in order to increase its weldability in the TTIr mode. For example, if clarifying agents are added to PP, the power density to make a weld greatly decreases and allows thicker samples to be welded with a given power density. It is also possible to weld thicker samples of amorphous PA compared to crystalline PA. In general, anything that reduces the amount of crystallinity or crystalline size increases a material's transmission of IR radiation.

While it is possible to increase a material's transmission by adding clarifying agents or reducing a material's degree of crystallinity, the design requirements of many applications prevent these changes. For example, clarifying agents tend to reduce a material's ultimate strength. Reducing a material crystallinity typically also reduces its solvent resistivity. If the design does allow material modifications, it has been shown that even with a weld time of 20 s it was not possible to weld PP with a thickness of 7 mm and power density of 5 W/cm in the TTIr mode. However, when clarifying agents were added, it was possible to weld the sample in 5 s [21].

In many applications, fillers such as glass fibers are added to a material to increase the material's ultimate strength. While these additives are advantageous with regard to design considerations, they tend to make the material less transparent and less weldable in the TTIr mode. For example, Fig. 12.34 shows the amount of power (power density) required to make a weld in 2 s with PA with increasing glass content.

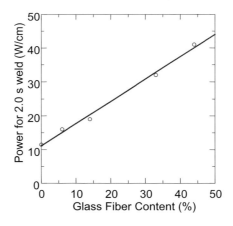

Figure 12.34 Typical power requirements for PA with various amounts of glass (3 mm thick sample)

As detailed in Section 12.3.2, the transmission through crystalline material decreases with thickness because of the material's inherent scattering characteristics. In contrast, the transmission through amorphous materials such as PS and PC tends to be thickness-independent. As seen in Fig. 12.35, the amount of power to weld PC and PS does not increase with thickness as it does with PA.

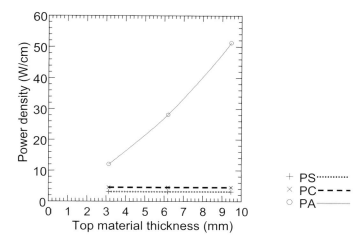

Figure 12.35 *Typical power requirement for selected materials as a function of sample thickness*

12.6.2   Effects of Dies and Pigments

It is important to understand that dies and pigments interact with IR radiation very differently. Dies tend to dissolve within a plastic and thus their particle size approaches the size of the molecular structure of the plastic. Because the particle size is so small, dies do not promote internal scattering of the IR radiation. In contrast, pigments tend to be inorganic and do not dissolve and their particle size depends on the manufacturing technique used. Thus, pigments tend to act as very small, randomly oriented mirrors, dispersed throughout the plastic promoting internal scattering. In fact, the amount of scatter can be so high that heating can not occur and thus TTIr welding is not possible.

In addition to their structural characteristics, the interaction of dies and pigments is also related to their color. For example, the color green is the result of strong absorption of red and high transmission or reflection of green. Thus, green materials tend to have a low transmission of IR radiation and often can be effectively heated with IR radiation. In contrast, red materials have very good transmission of IR radiation.

While carbon black is an effective additive for making a plastic absorb IR radiation, it is a black body and absorbs nearly the entire electromagnetic spectrum. Thus, the component will appear black and may even have some electrical properties depending on the amount of carbon black that was added. In some applications, it is important that both components appear black. In order to weld such an application with TTIr it is possible to add an infrared die to one component, to make it appear black but transmit IR radiation.

If it is critical that an application appears transparent, it is possible to add special dies that are relatively transparent in the visible spectrum but absorb in the near-IR spectrum. In this case, it is possible to weld two components, which both appear transparent with IR/laser welding, even with IR sources operating with wavelengths between 800 and 1,500 nm. Thus, part color does not necessarily limit the use of IR/laser [25].

12.7   Setup and Troubleshooting

Because machines vary greatly between modes of operation and manufacturer, the reader should consult the user's manuals or the manufacturer for assistance with setup and troubleshooting. In addition, due to the hazardous potential of IR radiation, only qualified personal should perform maintenance on these machines.

Various problems with IR/laser welding and their possible causes are listed in Table 12.5.

Table 12.5 Typical Problems and Their Possible Causes with IR/Laser Welding

Problem	Possible cause
Under weld	Power too low
	Time too short
	Diodes over aged
Weak weld	Time too short
	Clamp force too high
Porosity	Power too high
	Moisture in parts
	Clamp force too low
	Too much carbon black/absorbing additive
Too much flash	Clamp force too high
	Power too high
	Weld time too long
Un-welded zones	Head not aligned
	Failed laser(s)
	Dirt on pressure foot
	Part fit-up not proper

Problem	Possible cause
Inconsistency from cycle to cycle	Problem with laser cooling
Smoke/fumes	Power too high
Part marking (TTIr)	Power density too high
	Non-uniform additive distribution (such as SGF concentrated at surface)

If a laser has failed or dirt has built on a transparent pressure foot, a "burn pattern" on a black sheet of paper will help identify the problem. A way to produce a black paper is a photocopy made while the lid remained open on the copying machine. The result is a white and a black side. The black side should be placed in the fixture facing the IR/laser source so that it absorbs the IR radiation. After a short exposure to the IR radiation, the paper will heat and partially char, producing a burn pattern, see Figure 12.36. A failed laser/diode can usually only be replaced by the equipment manufacturer. If lamps are used, the reader should refer to the operating manual for further details on replacement procedures.

Sheets too close
to laser diodes   Proper burn pattern   Burn pattern with
                                                        failed diode

Figure 12.36 Typical burn patterns with continous illumination heating head

12.8   References

1. Grewell, D.A., Nijenhuis, W., *TTIr Welding of ALIPHATIC POLYKETONE*, ANTEC-00 Conference Proceedings, Society of Plastics Engineers, Brookfield, CT

2. Potente, H., Becker, F., *Weld Strength Behavior of Laser Butt Welds*, ANTEC-99 Conference Proceedings, Society of Plastics Engineers, Brookfield, CT

3. *U.S. Federal Laser Product Standard*: Title 21 of the Code of Federal Regulations; Part 1000 (parts: 1040.10 and 1040.11)

4. McComb, G., *The Laser Cookbook*, (1988) Tab Books Inc., Blue Ridge Summit, PA

5. Gray, H., *Chemical Bonds: An Introduction to Atomic and Molecular Structure*, (1973) W. A. Benjamin/ Cummings Publishing Company, Menlo Park, CA

6. Dwight, E.G., *American Institute of Physics Handbook*, (1957) McGraw-Hill, New York, NY

7. Gruenwald, G., *Plastics, How Structure Determines Properties*, (1993) Carl Hanser Publishers, Munich, Germany

8. Klein, H., Haberstroh, E., *Laser Beam Welding of Plastic Micro Parts*, ANTEC-99 Conference Proceedings, Society of Plastics Engineers, Brookfield, CT

9. Garton, A., *Infrared Spectroscopy of Polymer Blends, Composites and Surfaces*, (1992) Carl Hanser Publishers, Munich, Germany

10. Wolpert, D.,H., *Engineering Plastics*, (1995) **Vol. 2**, p.484, ASM Intern., Materials Park, OH

11. Kagan, V., *Optical Characteriz. of colored PAs for Laser Welding*, ANTEC-01 Conference Proceedings, Society of Plastics Engineers, Brookfield, CT

12. Grimm, B., Yeh, H., *Infrared Welding of Thermoplastics, Colored Pigments and Carbon Black Levels on Transmission of Infrared Radiation*, ANTEC-98 Conference Proceedings, Society of Plastics Engineers, Brookfield, CT

13. Potente, H., *The Effect of Moisture on Vibration Welds Made with PA*, ANTEC-92 Conference Proceedings, Society of Plastics Engineers, Brookfield, CT

14. Grewell, D.A., *Amplitude and Force Profiling: Studies in Ultrasonic Welding of Thermoplastics*, ANTEC-96 Conference Proceedings, Society of Plastics Engineers, Brookfield, CT

15. Stoke, V.K., *Vibration welding of thermoplastics. Part: Analysis of the welding process*, Polymer Engineering and Science, June 1988, **Vol. 28**, No.11, Society of Plastics Engineers, Brookfield, CT

16. Carslaw, H.S., Jaeger, J.C., *Conduction of Heat in Solids* (1959) Oxford University Press, London

17. Rosenthal, D., *The theory of moving source of heat and its application to metal treatment*, Transaction, ASME **43** (11): 849–866, 1946

18. Eagar, W., Tsai, N., *Temperature Fields Produced By Traveling Distributed Heat Source*, American Welding Society, Welding Journal, **Vol 162**, No 12, Dec 1983, PP. 346–355

19. Grewell, D., Lovett, D., US Patent 6064798, *Welding Method and Apparatus*, 1999

20. Grimm, R., US Patent 5843265, *Joining Method*, 1998

21. Grewell, D.A., *Applications with infrared welding of thermoplastics*, ANTEC-99 Conference Proceedings, Society of Plastics Engineers, Brookfield, CT

22. Jones, I. A., *Laser Welding of Plastic Films and Sheets,* Research Report, Oct. 1995, Num. 519/1995,TWI, Cambridge, UK

23. Yeh, H., Grimm, R., *Infrared Welding of Thermoplastics, Characterization of Transmission Behavior of Eleven Thermoplastics,* ANTEC-98 Conference Proceedings, Society of Plastics Engineers, Brookfield, CT

24. Grimm, R., *Through-Transmission Infrared Welding (TTIR) of Teflon®TFE(PTFE)* ANTEC-00 Conference Proceedings, Society of Plastics Engineers, Brookfield, CT

25. Jones, I., Taylor, N., *Use of Infrared Dies for Transmission Welding of Plastics,* ANTEC-00 Conference Proceedings, Society of Plastics Engineers, Brookfield, CT

13 Microwave Welding

Chung-Yuan Wu

13.1 Introduction

Microwaves are electromagnetic waves containing time-varying electric and magnetic fields. Microwave frequencies range from 300 MHz to 300 GHz, placing them between the far-infrared and the FM broadcast radio band in the electromagnetic spectrum. Modern high-power microwave ovens are operated at 2450 MHz according to the restrictions of the Federal Communication Committee (FCC) in the USA. Microwave heating of materials is based on the material's absorption, where some materials absorb the magnetic field, some absorb the electric field, and some absorb both. The major applications are in drying, thawing, cooking, and heating, while radar and telecommunication are other fields of application using microwaves. Industries with a wide range of microwave applications include the paper, printing, leather, textile, wood, plywood, ceramics, rubber, and plastic industries. In the paper industry, microwaves are not only used for the evaporation of water but also for drying of coatings and glue on the paper. The major advantage of using microwaves is that it raises the bulk material temperature volumetrically instead of by conduction from the surface such as conventional oven heating. It is especially useful for heating low thermal conductivity materials.

The use of microwave radiation for welding thermoplastic materials is a relatively new technology, and therefore it is not commonly used in industry. However, because microwave welding has several unique benefits, a fair amount of research has been conducted and it is anticipated that its popularity will grow. One of the unique benefits is that it is possible to weld internal walls that may not be accessible by other welding techniques, because most virgin plastics are transparent to the microwave radiation.

Typically, microwave welding of thermoplastics is achieved by placing an absorbing material into the bond line to enhance microwave absorption. During the process, the gasket (insert) acts as a consumable and is squeezed out from the bond line. If the gasket/insert remains in the bond line at the end of the cycle, it can be reheated and used to disassemble the joint.

13.2   Process Description

Microwave welding of thermoplastics involves the following steps:

1. *Application of gasket.* A consumable component is applied to the weld interface. This is currently a manual process, but it is envisioned that it can be automated with robotics or even with pad-printing techniques or insert molding

2. *Clamping of parts.* The components are clamped with a specially designed fixture. This assures that the faying surfaces are in intimate contact during the welding cycle. The fixture must be designed to either fully reflect or transmit the microwave radiation so that it does not heat during the welding cycle.

3. *Microwave heating.* The parts are exposed to microwave radiation until the bond line is fully melted.

4. *Hold phase.* The molten material is allowed to solidify and heal the weld. Pressure is usually required during this phase; any consumable remaining is squeezed out of the bond line to achieve high weld strength.

Table 13.1 lists the typical cycle times required for microwave welding.

Table 13.1 Typical Microwave Welding Cycle

Step	Typical time [s]
Loading parts	3 to 10
Press actuation	1 to 2
Microwave radiation	5 to 20
Hold phase	3 to 10
Part removal	5 to 10

13.3   Physics of the Process

13.3.1   Microwave Heating

Most of the applications involving microwave heating of non-conducting materials rely on dielectric losses of the material being heated. The heating is based on the internal charges (charges based on chemical structure) and their freedom of movement to align themselves with an applied alternating electric field. In general, there are four types of polarization, namely:

*   Electronic,
*   Atomic,
*   Dipolar and
*   Interfacial (Maxwell-Wagner) polar [1].

Electronic polarization occurs when the electrons around the nuclei are displaced generally to one side under the influence of an external electric field. Atomic polarization is caused by the movement of atomic nuclei due to uneven distribution of the charges in the molecule. Dipolar polarization is caused by the presence of permanent dipoles in the materials. Maxwell-Wagner polarization occurs in the charges that were built at interfaces in a heterogeneous system. For a conducting material, D.C. conductivity has to be taken into account in the microwave heating.

When considering the heating of a material, it is convenient to refer the energy that is converted to heat as the "loss" energy. Thus, a material with a high loss has a relatively high ability to convert the electrical energy into heat. There are two major losses associated with plastics in a microwave field. The first, Joule losses, occurs when a material has a finite conductivity (σ) and there is resistive heating. The second loss is due to molecular vibrations induced by the alternating electric field (polarization). In order to account for both losses, an effective loss factor (ε''_{eff}) is used. The measure of total losses is often defined in terms of the effective loss tangent $(\tan \delta_{eff})$ as defined in Eq. 13.1 [1].

$$\tan \delta_{eff} = \frac{\varepsilon''_{eff}}{\varepsilon'} \tag{13.1}$$

Where ε' is the real part of the dielectric constant, and the effective loss factor is defined in Eq. 13.2.

$$\varepsilon''_{eff}(\omega) = \varepsilon''_c(\omega) + \varepsilon''_e(\omega) + \varepsilon''_a(\omega) + \varepsilon''_{MW}(\omega) + \frac{\sigma}{\varepsilon_0 \omega} = \varepsilon''(\omega) + \frac{\sigma}{\varepsilon_0 \omega} \tag{13.2}$$

where the subscripts d, e, a, MW, represent dipolar, electronic, atomic, and Maxwell-Wagner, respectively, ω is the angular frequency, ε_0 is the permittivity of free space, and σ is conductivity.

When exposed to an external alternating field, a material's heating is proportional to its effective loss tangent. However, because the loss tangent is a function of temperature, it is difficult to predict the amount of heating accurately. In addition, it is not only a function of temperature but also a function of the time the material is exposed to the elevated temperature. Therefore, the heating calculations can become very complicated. A general and useful formula to calculate the average power dissipation (P_{ave}) in the material under high frequency radiation is Eq. (13.3) [1].

$$P_{ave} = \omega \varepsilon_0 \varepsilon''_{eff} E^2_{rms} V \tag{13.3}$$

Where E_{rms} is the root mean square value of the electric field strength (volt/meter) inside the material, and V is the volume of the material. It should be noted that this expression includes dielectric losses only and does not include magnetic losses. If a material shows a magnetic loss then the total losses are defined as in Eq. 13.4.:

$$P_{ave} = \omega \varepsilon_0 \varepsilon''_{eff} E^2_{rms} V + \omega \mu_0 \mu''_{eff} H^2_{rms} V \tag{13.4}$$

Where μ_0 is the permeability of free space, μ''_{eff} is the effective magnetic loss factor, and H_{rms} is the magnetic field strength inside the material. Once the material properties become temperature dependent, the temperature prediction again becomes difficult.

As detailed in Chapter 2, it is critical to heat the faying surfaces in order to promote molecular diffusion and to produce a weld. Thus, it is important for the material to have a relatively high loss factor (such as PVC and Nylon) in order for microwave welding to work. However, because of the volumetric heating with microwave radiation, the peak temperature always occurs at the center of the parts, which does not usually coincide with the location of the joint. Thus, it is difficult to limit heating to the faying surfaces. To limit the heating to the faying surface, it is common practice to place a microwave sensitive material (absorber) such as a conductive polymer at the weld interface. A composite gasket (or film) is used for handling purposes as a preferred form. There are several ways to fabricate the composite, such as extrusion, compression molding, ultrasonic molding, and microwave molding.

13.3.2   Conductive Polymers as Microwave Absorbers

During the past 25 years, new intrinsically electrically conductive polymers, such as polythiophene, polypyrrole, and polyaniline, have been intensively studied. The pioneers in this field, Heeger, MacDiarmid, and Shirakawa, won the Nobel Prize in Chemistry in 2000. These materials have a unique combination of mechanical and electrical properties making them very useful for welding. Because, polyaniline is relatively inexpensive, is easy to synthesize and process, and it is stable at room temperature [2], it has been the primary material studied for microwave welding.

Similar to semiconductors, polyaniline conducts electricity through doping that creates partially filled bands through which free moving electrons conduct electricity. The electrical properties of polyaniline are a function of frequency, temperature, morphology, and doping level. Stretched films or fibers that have a higher level of crystallinity have a higher electrical conductivity than unstretched films and powders. The electrical conductivity of polyaniline can be varied from that of an insulator (10^{-10} s/cm) to that of a conductor (10^4 s/cm), depending on the processing and manufacturing techniques. Figure 13.1 shows the chemical structure of polyaniline for (a) fully reduced form (LEB), (b) 50% oxidized form (EB), (c) fully oxidized form (PNB) and (d) emeraldine hydrochloride salt polymer (ES) [3].

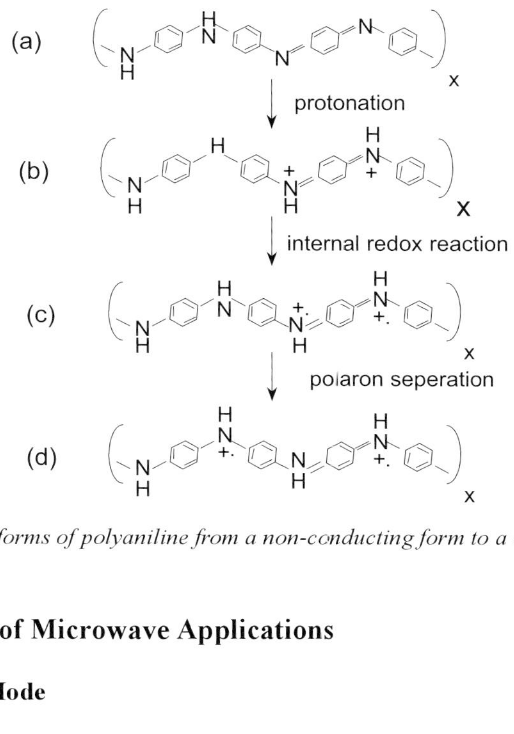

Figure 13.1 Different forms of polyaniline from a non-conducting form to a conducting form [3]

13.3.3   Modes of Microwave Applications

13.3.3.1   Multi-Mode

The most widely used microwave oven (domestic kitchen microwave oven) is a multi-mode microwave cavity. The dimensions, operating frequency (2.45 GHz in the United States), and dielectric constant of the material inside the cavity determine the modes of operation. When the cavity is partially filled with an irregularly shaped dielectric material, the number of modes and wave pattern becomes very complicated and difficult to calculate analytically. Finite element analysis and finite difference time domain method can be used to calculate the field distribution. In many cases a multi-mode field is desired since it can often result in relatively uniform volumetric heating.

Multi-mode microwave welding of thermoplastics has been demonstrated to result in relatively high joint strength; however, the heating time is relatively long compared to some of the other welding techniques, such as ultrasonic welding. There are several methods that can reduce the heating time such as using a higher microwave power, using more lossy absorbers, or increasing the power transfer efficiency. The peak heat generation rate of a

60% PANI composite (6.35 mm x 6.35 mm x 0,5 mm) in a 600 W multi-mode microwave is in the order of 2×10^8 W/m^3 as shown in Fig. 13.2; thus, the peak power absorption is only about 4 watts.

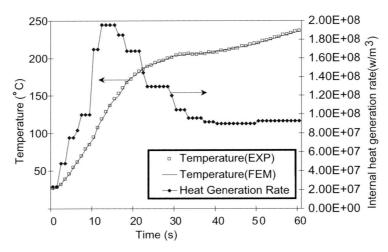

Multi-Mode Microwave Welding 600Watts, 60% -0.5mm

Figure 13.2 Experimental and FEM predicted internal heat generation rate during multi-mode heating

13.3.3.2   Single-Mode

The source for a single-mode microwave oven is the same as in the multi-mode. The heating cavity dimensions determine the mode generated during welding. Therefore, a waveguide is usually used to produce a single mode. For the 2450 MHz operating frequency, an S band rectangular waveguide is used to produce a single mode that has a cross section of 34 mm x 72 mm. In this case, the waveguide also serves as a heating chamber. In general, a single mode microwave system is one in which the electric field is well defined and controlled, making it possible to orient the sample (parallel to the electric field) to maximize the heating.

A basic single mode microwave heating system consisting of a microwave power source, a controller, a 3-port circulator, a 4-stub tuner, an applicator, two dual power meters, and two dummy loads is shown in Fig. 13.3. Welding is done in the double slotted applicator (waveguide) that is connected to the tuner.

The instantaneous electric field distribution for a single mode, TE$_{10}$ mode, can be calculated by solving the Maxwell's equations with appropriate boundary conditions as shown in Figure 13.4.

The wave operated in Fig 13.3 is defined as traveling wave because the second dummy load absorbs the incoming wave without reflection.

In some cases it is important to generate traveling waves, where the waves are traveling down the waveguide result in uniform electric field strength in the waveguide. In general, a traveling wave is used for the heating of large structures and in continuous processing.

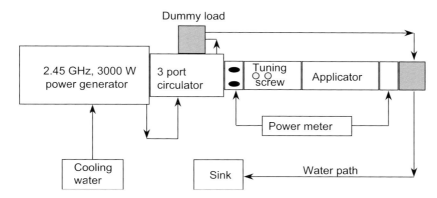

Figure 13.3 *Schematic representation of a single mode microwave welding system [4]*

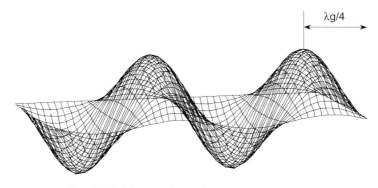

For air filled S band, f=2.45GHz,   (λg/4)=5.8mm

Figure 13.4 *Three-dimensional plot of the electric field distribution for a TE_{10} mode*

Another basic operation in a single-mode microwave system is the standing wave pattern that can be generated by removing the second dummy load in Fig. 13.3 and replacing it by a solid plate (shorted plate). The constructive interference between the forward and the reflected wave results in higher electric field strength. Figure 13.5 shows the electric field strength measured at 100 W input power in the traveling and standing wave conditions. It is important to note that the electric field strength is a half-sinusoidal variation. The standing wave pattern has uneven field distribution that results in non-uniform heating. However, it is beneficial for welding of thermoplastics to use a conducting composite because it enables maximizing the field strength at the interface thereby producing very localized heating

In order to maximize heating in single-mode microwave welding, the gasket should be placed at a quarter wavelength from the shorted plate. It should be noted that there are three possible orientations for the gasket, as shown in Fig. 13.6. The bottom diagram represents the interaction between gasket and applied electric field. If the sample is placed as in Orientation 1 in Fig. 13.6, the resulting electric field within the gasket is relatively small and the heating is limited, because of the large dielectric constant of the gasket. Moreover, depending on sample size and shape, Orientation 3 may require openings on the sidewalls that generate significant leakage. Therefore, Orientation 2 is recommended for welding.

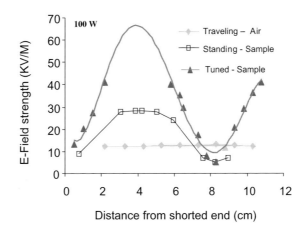

Figure 13.5 Measured electric field strength at 100 W standing wave

In order to maximize heating, it is recommended that the electric field in the waveguide be measured/mapped in order to assure the samples are placed at the maximum field. Because the sample and fixtures affect the electric field, it is important to map the field with the samples and fixtures placed in the waveguide. For example, Fig. 13.7 shows the possible effect of fixture and sample on the electric field distribution.

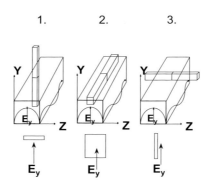

Figure 13.6 Possible microwave welding orientations in waveguide

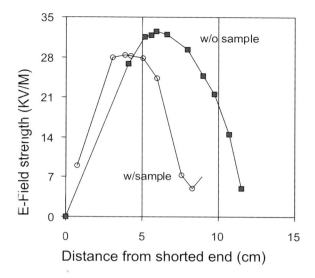

Figure 13.7 Electric field strength distribution in waveguide with sample and fixture

13.3.4   Heat Generation Rate

The gasket temperature in microwave welding is similar to the hot plate temperature in hot plate welding; therefore, understanding the temperature history in the gasket is important in microwave welding. Figure 13.8 shows the temperature history of a 50%-0.5 mm and a 60%-0.5 mm gasket during microwave heating.

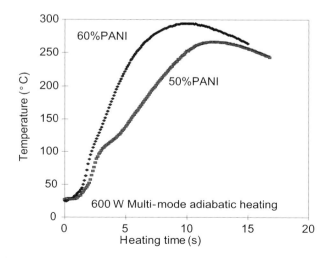

Figure 13.8 Temperature history in 50% and 60% polyaniline gaskets in multi-mode microwave

To determine the initial power dissipation in the gasket, it is necessary to determine the electric field distribution in the material. Assuming that the conductive term dominates the heating and the electric field in the gasket is uniform with constant magnitude (gasket is small compared to the wavelength and skin depth), the average power dissipated in the gasket is defined in Eq. 13.5.

$$P_{avg} = \sigma E_{rms}^2 V \qquad (13.5)$$

In Eq. 13.5, P_{avg} is the average power dissipated, E_{rms} is the root mean square value of the electric field strength in the gasket, and V is the volume of the gasket. It is important to note that the electric field strength in the gasket is related to the conductivity of the gasket. Therefore, increasing the conductivity of the gasket may not necessarily increase the power dissipated. For example, a perfect conductor reflects the microwave and the internal electric field strength is zero, resulting in no heating. Unfortunately, determining the internal field strength in the gasket is very complicated, since it is a multi-mode microwave cavity. Therefore, the field strength and the internal heat generation rate in the gasket is usually estimated using calorimetry or adiabatic heating of the gasket while measuring the temperature rise. By using 1-D adiabatic internal heat generation as defined in Eq. 13.6:

$$\rho C_p \frac{\partial T}{\partial t} = \dot{Q} = \frac{P}{V} = \sigma E_{rms}^2 \qquad (13.6)$$

and rearranging Eq. 13.6, E_{rms} can be defined as seen in Eq. 13.7.

$$E_{rms} = \sqrt{\frac{\rho C_p}{\sigma} \frac{dT}{dt}} \qquad (13.7)$$

where ρ is the density, C_p is the specific heat, σ is conductivity, T is temperature, and t is time. Using Eq. 13.7 requires the definition of material properties such as density, specific heat, and composite conductivity. For simplicity reason it can be assumed that the density and conductivity are temperature-independent. The other material property that must be defined in order to model the heating is the specific heat (C_p) of the material. For reference purposes, Fig. 13.9 shows the specific heat of pure Polyaniline-HCl doped powder and HDPE as a function of temperature. It is seen that the peak value of the C_p of HDPE occurs at that the T_m at 132 °C. Using the rule of mixture, the value for a 50% PANI composite was 2511.5 J/Kg °C and for a 60% PANI composite was 2464 J/Kg °C. The complex permittivity of the gasket can be measured at room temperature using a transmission line technique. For reference purposes, at 2.45 GHz the conductivity was 8.13 s/m for 60% PANI composite and 3.34 s/m for 50% PANI composite. The loss tangent at 2.4 GHz for 60% and 50% PANI composite was 1.23 and 0.69, respectively. The heating rate (dT/dt) can be obtained by direct temperature measurement. Table 13.2 shows the estimated electric field strength in the gasket at 60 °C. It was found that higher conductivity results in a lower electric field in the material. This is expected because the electric field is reflected more from a higher conductive surface. The initial heat generation rate can be calculated from Eq. 13.6. The average initial heat generation rate for a 50% 0.5mm thick PANI composite was 4.0×10^7 w/m^3 while for a 60% 0.5mm thick PANI composite it was 3.52×10^7 w/m^3 [5].

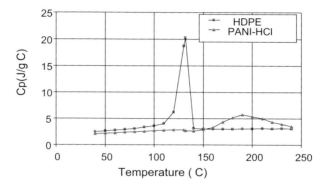

Figure 13.9 Specific heat of HDPE and pure PANI-HCl measured from DSC

Table 13.2 Estimated Electric Field Strength in Gasket at 60 °C from Adiabatic Heating

Sample No.	Field strength (volt/meter) 60% PANI-0.5mm thick	Field strength (volt/meter) 50% PANI-0.5mm thick
1	2090	3360
2	1920	3250
3	2230	3750
Average field strength	2080	3460
Initial heat generation rate	3.52×10^7 (w/m^3)	4.00×10^7 (w/m^3)

To extend the modeling through the entire welding cycle it is necessary to use finite element method (FEM) in conjunction with temperature measurements. Considering the one-dimensional heat transfer equation (Eq.13.8):

$$\rho C_p \frac{\partial T}{\partial t} = \lambda \frac{\partial^2 T}{\partial x^2} + \dot{Q}$$
(13.8)

where λ is thermal conductivity. The internal heat generation rate $\dot{Q}$ of the gasket can be determined by matching the temperature history output form FEM to the experimentally measured temperature history at the same location. Figure 13.10 shows typical temperature histories during welding for 60% PANI-0.5mm and 50% PANI-1mm gaskets in multi-mode microwave welding. The heat generation rate in microwave welding for different gaskets can be estimated using a one-dimensional finite element heat conduction model (ignoring convection losses at surrounding). This heat generation rate can be found by changing the value of the heat generation rate with respect to time in the FEM analysis to obtain a temperature history at a specific point that matches the experimental measurement at the same point. Figure 13.11 shows the experimental and computer simulated temperature histories and the estimated heat generation rate of the 50% PANI-1 mm gasket. In this case, the maximum heat generation rate in the 50% PANI gasket is 7.7×10^7 W/m^3. The decrease in the heat generation rate for elevated temperatures is caused by the changes in the electrical properties of the composite gasket. This change is caused by elimination of the HCl on the amino group and by chlorinating of the aromatic ring in the polyaniline. In comparison to Fig. 13.2, the maximum heat generation rate for 60%-0.5mm gasket was 1.96×10^8 w/m^3 that is higher than the 50%-1mm gasket. It is interesting to note that the 50% PANI-1mm thick gasket contains more polyaniline powder compared to the 60% PANI-0.5mm gasket, but the latter produced a higher temperature rise rate and higher maximum temperatures. There are many advantages modeling the process, such as prediction of molten layer thickness, determination of the optimal PANI-HCl content in the gasket and the thickness of the gasket.

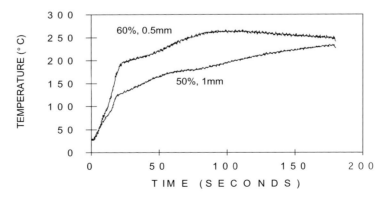

Figure 13.10 *Temperature histories at the interface between the HDPE bar and conducting composite using multi-mode*

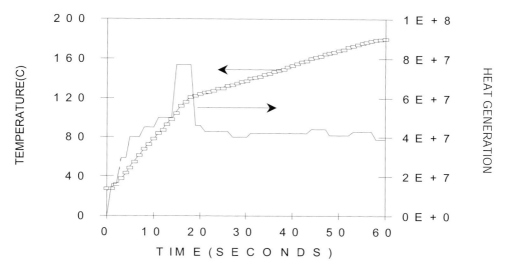

Figure 13.11 *Experimental and FEM predicted internal heat generation rate during welding for 50% PANI, 1mm thick gasket*

13.4 Equipment Description

13.4.1 Equipment Components

There are three major components in a microwave welding system:

Generator/Power Supply: Most of the microwave power source for most systems is a magnetron tube. A magnetron is a vacuum resonator (see Fig. 13.12) [5]. It contains a hollow cylindrical anode with a cathode that is heated at the center of the structure. The slots between the anode and the cathode act as the resonant cavities that are separated by vanes. A constant voltage drop is applied between the anode and the cathode to create an attraction force for the electrons released from the heated cathode. A magnetic field is applied parallel to the cathode axis. Due to Lorentz's Law, $F = q \ (v \ x \ B)$, where F is the force, q is the charge, v is velocity, and B is magnetic field strength, a circular force around the cathode is created and the moving electrons form an electron cloud. During resonance, the electric field strength at the vane walls is zero; however, by transmission line theory, the open tips will have maximum electric field strength. Due to the π mode resonance, the adjacent cavity has a phase difference of 180°; thus, a strong electric field occurs between the vanes. These fields are altered with respect to time between vanes as shown in Fig. 13.12. Therefore, electron clouds are decelerated by the opposite direction to the electric field and they fall to the anodes. On the other hand, if the field direction is the same as the electron motion direction, electrons are accelerated toward the cathode until they decelerate and give up their energy to the induced microwave field [6]. The microwave energy can be coupled by coaxial cable or by a waveguide. Once the waves are introduced into the cavity, they propagate in all direction and interfere with each other.

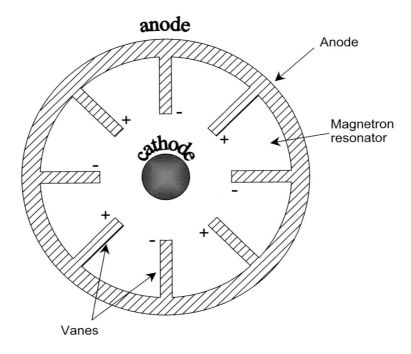

Figure 13.12 *Schematic representation of a typical magnetron cavity*

Controls: The controls establish an interface between the machine and the operator monitoring the system. Because currently there is no commercial microwave welding equipment for thermoplastics, most microwave equipment used for welding of plastics is for drying and cooking. Thus, the controls are usually limited to a pre-selected weld time and power level.

Fixture: This component holds the parts being welded during the weld cycle. Because the welding fixture must be exposed to the microwave radiation, it is usually fabricated from non-polar plastics, such as PP, PE and Teflon. If high clamping forces are required, pneumatic cylinders can be used but they must be placed outside the cavity. The cylinders can usually be coupled to the fixtures through a hole in the cavity. It is important to note that the hole can result in leakage. This leakage not only presents an issue with the system efficiency but also a safety risk.

13.4.2   Applications

Currently there are no commercial applications for welding of thermoplastics, but because of some of the benefits of the process it is believed that future development will make this process commonly accepted by the industry. For example, microwave welding can weld internal hidden walls therefore applications such as batteries or liquid reservoirs may be welded with microwave welding in the future.

Welding of large and/or complex structures with multiple joints is impossible with other plastics welding methods. However, microwave welding of these components may prove viable. Large microwave cavities and high power sources are already available and it may be easier to weld large structures using conductive gaskets compared to other conventional welding techniques.

13.5   Material Weldability

Because of the limited number of applications documented there is also limited information on the weldability of materials for microwave welding. While any thermoplastic with a low dielectric loss should be weldable with microwave welding, assuming a gasket insert approach, only HDPE and Nylon have been studied in detail. The following section provides information on the welding of HDPE and Nylon with an additional section on the general weldability of several additional materials.

13.5.1   Multi-Mode Microwave Welding of HDPE

The parameters that affect the final joint strength are PANI doping level, concentration of PANI in the gasket, welding time, welding pressure, and gasket thickness. The PANI doping level affects the intrinsic conductivity and loss tangent of the polyaniline powder that affects the properties of the composite gasket. The welding parameters such as welding time and welding pressure are the most important factors. Moreover, material parameters such as the gasket thickness and percentage of the polyaniline in the gasket are can also affect weld strength.

Figure 13.13 shows the effect of welding time and PANI concentration on joint strength for HDPE. The average strength of a 50% PANI gasket is only 12.34 Mpa, which is close to the joint strength made at 120 s. For welding times longer than 120 s, all samples had better joint strength than the strength of the gasket, because the molten gasket breaks under pressure and squeezes out from the interface.

Increasing the gasket thickness increases the PANI content in the gasket resulting in more heating while using a constant pressuring method. As shown in Fig. 13.14, weld strength is proportional to gasket thickness, because it takes longer to squeeze out thicker gaskets, which results in more heating. The effect of welding pressure on joining strength is shown in Fig. 13.15. It is seen that higher pressure results in higher joint strength and it also shortens the welding time needed to reach a certain strength. Figure 13.16 shows the photographs of (a) an intact gasket and (b) a squeezed out gasket.

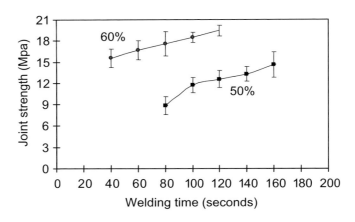

Figure 13.13 *Effect of PANI concentration on joint strength with 0.3 MPa joint pressure using multi-mode and constant pressuring method*

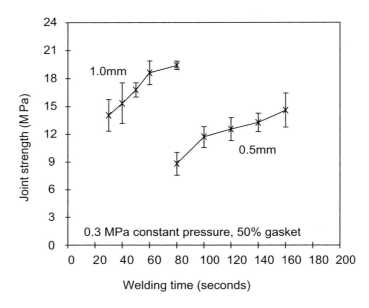

Figure 13.14 *Effect of gasket thickness on joint strength for a 50% PANI gasket with 0.3MPa joint pressure using multi-mode and constant pressuring method*

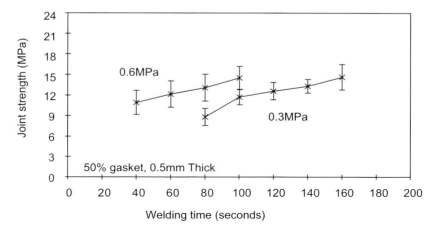

Figure 13.15 Effect of welding pressure on joint strength for a 50% PANI 0.5mm thick gasket using multi-mode and constant pressuring method

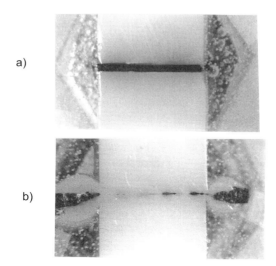

Figure 13.16 Microwave welded joints; (a) intact gasket, and (b) squeezed out gasket

It is possible to reduce cycle time and increase weld strength by using a force-profiling technique. By applying a relative low weld force during the heating cycle it is possible to retain the gasket in the joint and produce more melt. Once the entire interface is melted and the melt layer becomes relatively thick, high pressure can be applied. The high pressure squeezes the gasket out and produces intimate contact resulting in a relatively strong weld. Figure 13.17 shows the comparison of constant heating and the post heating pressure. It is clear that the post heating pressure is the preferred welding method [7].

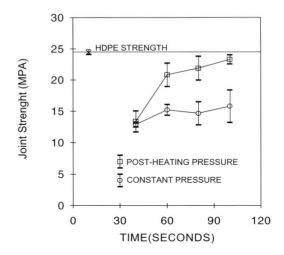

Figure 13.17 *Comparison of constant heating and the post heating pressure*

13.5.2   Single-Mode Microwave Welding of HDPE

As mentioned in the previously section, a single-mode microwave system provides high power efficiencies. Figure 13.18 shows the effect of heating time and power on joint strength for single-mode microwave welding using a 60% PANI 0.5 mm gasket for welding HDPE. As expected, 2400 W provide faster heating and a shorter welding cycle. It is important to note that at 15 s, the joint strength is 26.50±0.74 MPa, which is 96% of the bulk material strength (27.58±0.32 MPa). Figure 13.19 shows the effect of PANI concentration in the composite on joint strength. It is seen that using 50% PANI loading gasket provides a good joint quality using 20 s of heating time. The joint strength can be further increased either by increasing the heating time or increasing the joining pressure. Using a 30% PANI gasket with 20 s of heating time results in a weld strength of 55% of the HDPE strength.

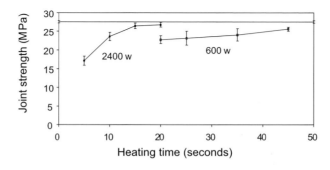

Figure 13.18 *Effect of heating time and power on joint strength using 60% PANI 0.5 mm gasket in single-mode*

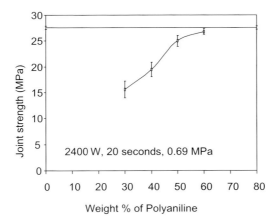

Figure 13.19 Effect of polyaniline concentration on joint strength

There are two more techniques that can further reduce the welding time: using a more conductive polymer and using impedance matching technique [8]. The impedance matching technique or resonance cavity is designed to generate higher electric field strength in the welding waveguide. An example of the electric field strength measurement is shown in Fig. 13.5. The electric field strength of an impedance matching system, i.e., the tuned system in Fig. 13.5, increased many times than the traveling wave. The use of more conducting polymer together with the impedance matching technique not only reduces the weld time but also reduces the polyaniline concentration required to achieve high quality welds. Figure 13.20 shows the welding results using a 750 W impedance matching system and 10% Version® polyaniline.

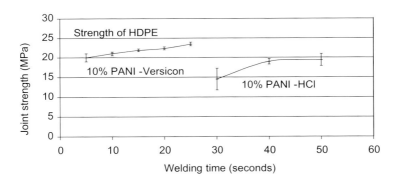

750W, Tuned, 10% PANI + HDPE = 0.5 mm

Figure 13.20 Weld strength as a function of weld time with HDPE and various compositions of PANI

13.5.3   Single-Mode Microwave Welding of Nylon

Welding Nylon is different from welding HDPE because Nylon contains polar groups and thus experiences bulk or volumetric heating. The polar groups heat up under microwave radiation without the need for a heating element. Thus, using an absorbing material is crucial in order to prevent over-heating, thermal run-away, due to the fact that polyaniline loses its conductivity and heating as the starts to Nylon melts. Film thickness, size, and polyaniline concentration can greatly affect weld quality. Figure 13.21 shows the adiabatic heating of PANI Nylon films. The PANI concentration in the films varied between 0.8% and 50%. The films with the best heating properties contained 1.6% and 7.7% PANI. The 0.8% and 50% PANI films resulted in a low heating rate and low temperatures. When the power is low, the temperature at the interface is close to the bulk temperature; however, at higher power levels, the temperature rises dramatically when the bulk temperature exceeds 100 °C. This phenomenon causes thermal run away. It is crucial to control the heating time in microwave welding of polar materials. In addition, Nylon absorbs moisture, thus all samples have to be dried before welding to reduce heating of water. Figure 13.22 shows the effect of welding time on weld strength using 2000 W standing wave, 4.57%–0.05 mm film and 1 MPa welding pressure. The weld strength can be further increased either by increasing the weld time or by using different PANI concentration. The maximum joint strength was 71.45±2.64 Mpa, which is 97% of the base Nylon 6/6 strength [9].

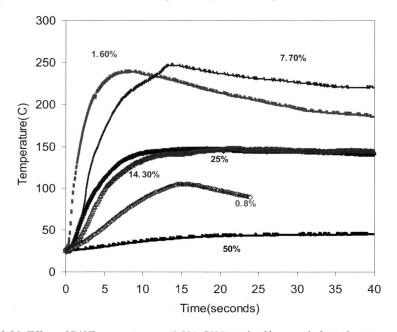

Figure 13.21 *Effect of PANI concentration (0.8%–50%) in the film on adiabatic heating*

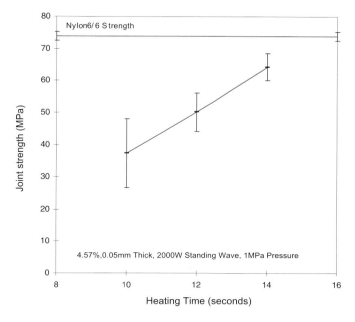

Figure 13.22 *Effect of welding time on joint strength*

13.5.4   Additional Materials

Welding of other thermoplastic materials such as PP, PC, PETG is also possible. Figure 13.23 shows the maximum joint strength for some other polymers using a 60% 0.5 mm PANI + 40% HDPE gasket and a multi-mode microwave system. The joint strength achieved is relatively high considering that the gaskets were molded from HDPE and not the polymer that was welded. There is also considerable latitude in the processing conditions to optimize the joint strength.

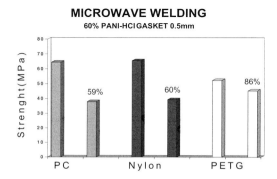

Figure 13.23 *Maximum joint strength for PC, Nylon and PETG using HDPE conducting composite*

13.6   References

1. A. C. Metaxas and R. J. Meredith, Industrial Microwave Heating, Peter Peregrinus, 1983

2. M.K. Traore, W. T. K. Stevenson. B.J. McCormick, R.C. Dorey, S. Wen and D. Meyers, *Synth. Met.*, **40**, 1991, pp. 137

3. M. Aldissi (ed.), Intrinsically Conducting Polymer: An Emerging Technology, Kluwer Academic Publishers. 1993, pp. 165–178

4. C. Y. Wu and A. Benatar, "Single Mode Microwave Welding of HDPE Using Conductive Polyaniline Composites", SPE, ANTEC 1995, p. 1244

5. H. E. Thomas, Handbook of Microwave Techniques and Equipment, Prentice-Hall, 1972

6. R. E. Collin, Foundations for Microwave Engineering, 2nd edition, McGraw-Hill, 1992

7. C.Y. Wu and A. Benatar, "Microwave Welding of High Density Polyethylene using Instrinsically Conductive Polyaniline", *Journal of Polymer Eng. and Science*, **Vol. 37**, No.4, p. 738, April 1997

8. C.Y. Wu, S. Staicovici, and A. Benatar, "Microwave welding of HDPE using Impedance Matching System", *Journal of Reinforced Plastics and Composites*, 1998, p. 27

9. C.Y. Wu, S. Staicovici, and A. Benatar, "Single Mode Microwave Welding of Nylon 6/6 Using Conductive Polyaniline Film", SPE, ANTEC 1996, p. 1250

14   Guidelines for Process Selection

Joon Park and David Grewell

14.1   Introduction

While many applications can be joined by more than one process, there are usually only one or two processes that are best suited for any given application. For example, when sealing PVC medical bags, it is possible to use ultrasonic welding, impulse welding, or RF welding. However, ultrasonic welding may not be able to produce the uniformity of the seal around the entire component and may also produce particulates that render the final component useless. Impulse welding may not be able to weld the complex geometries that may exist. However, RF welding can easily meet all the requirements and is the best process for the application and manufacturing requirements. Therefore, the selection of the most appropriate process for components should be made very carefully to assure that the manufacturing process is robust and the final component meets all requirements.

The welding process affects a wide range of a company's costs, including:

- Equipment capital cost
- Productivity
- Scrap rate
- Floor space
- Labor requirements
- Warranty liability of the products.

Selection of the appropriate process for assembly is not always straightforward. It must be based on many process requirement considerations such as:

- Material characteristics
- Part size
- Joint geometry
- Manufacturing and production requirements
- Cost analysis

Often the process selection is complicated by the fact that many of these factors are unknown or not well defined. For example, during a design phase of a new product, the material selection may change several times, while at the same time the joint design must be properly matched for the process. Thus, it is critical that all designs are based on concurrent engineering practices in order to minimize such problems.

Only after defining the process requirements as best as possible can a process be selected. Consulting an expert is often very beneficial.

In the following sections advantages and disadvantages for each of the processes are reviewed along with examples of typical applications for each process. The reader is encouraged to refer to the process chapters for more detailed descriptions. Since process selection is often a complex and ambiguous task, the following sections only provide a general overview.

Following the overview of each of the welding processes, the process requirements for manufacturing are detailed.

14.2   Summary of Process Characteristics

14.2.1   Hot Plate Welding

The hot plate welding process utilizes a heated platen to heat and melt thermoplastic material at the joint interface. In contact hot plate welding, the parts are brought into physical contact with the hot plate and heating occurs by conduction. In non-contact hot plate welding, the part surface is brought close to the hot place (gap of 1 to 3 mm) and heating occurs by a combination of both convection and radiation. Non-contact welding is usually used with high temperature thermoplastics or low viscosity thermoplastics, such as acetal. In both cases, after a molten layer is formed on the surface, the hot plate is retracted and the parts are pressed together.

Advantages:

- Well suited for large and complex parts
- More tolerant on material compatibility
- Wide operating windows
- Larger Heat Affected Zone (HAZ) can benefit the residual stress
- Flash is usually smooth and fully attached (no particulates)

Disadvantages:

- High utility cost
- Slow process (10 to 50 s)

- Sticking of residual plastic to hot plate and possible contamination
- Highly visible flash
- More difficult and expensive for tool changes

14.2.2   Hot Gas and Extrusion Welding

In hot gas welding, a hot gas gun is used to heat a welding (filler) rod and the surfaces to be joined. The operator applies pressure to the weld rod while proceeding along the joint interface. Usually, a taper or V-groove joint design is cut into the plates to accept the molten polymer from the welding rod. As detailed in Chapter 4, there is a wide range of possible joint designs. Often this technique is compared to oxy-acetalene welding of metals. Once the molten interface and filler material are allowed to cool and solidify, the weld is complete. Extrusion welding is very similar to hot gas welding, but instead of melting the filler with hot air, the filler material is melted in a screw design and extruded into the joint design. The screw design is similar to what is used in injection molding and extrusion machines, however, on a smaller scale.

Advantages:

- Flexible
- Low equipment cost
- Simple joint design
- Ideal for short runs
- Amenable to large parts

Disadvantages:

- Slow (0.5 to 1.0 cm of linear weld/s)
- Weld quality is dependent on operator skill
- Polymer degradation is possible

14.2.3   Implant Induction Welding

In induction welding, an alternating magnetic field is applied to a gasket placed in the joint. This gasket is filled with a ferromagnetic material that heats because of the magnetic field. Typically, the gasket is molded from either the same polymer as the parts being joined or from a polymer that is compatible with the part material. During heating of the gasket, the part surfaces melt, and after a pre-selected time, the field is turned off and the parts are cooled under pressure to produce a weld. It is important to note that the gasket becomes a permanent part of the assembly and is a consumable. That is to say the gasket is not reusable from one cycle to the next.

Advantages:

- Medium speed (5–10 s)
- Highly controlled heating and cooling is possible
- It can be automated
- Low residual stress
- Non-contact welding
- Able to weld un-supported internal walls

Disadvantages:

- Limited by the distance between induction coil and implant gasket
- Induction coil design is important
- Impedance matching between coil and gasket can be critical
- Consumable gasket can add to the cost of the process

14.2.4   Implant Resistance Welding

Implant resistance welding is primarily used in the gas/water distribution industry to join pipes and fittings. The process works by passing electrical current through a wire that is placed at the joint interface and relying on joule (resistive) heating of the wire to promote heating and melting of the interfaces. Usually the parts are designed so that intimate contact of the faying surfaces is produced by the thermal expansion of the heated material.

Advantages:

- No displacement/collapse is required during welding
- Equipment is relatively simple (DC power supply is sufficient)
- Can weld relatively large parts
- Relatively fast (10–30 s)

Disadvantages:

- Socket or saddle component with wire imbedded at faying surface is consumable
- Remaining wire can act as stress concentration point

14.2.5   Ultrasonic Welding

In ultrasonic welding, high frequency and low amplitude mechanical vibrations are applied to the parts. These vibrations result in cyclical deformation of molded triangular protrusions at the faying surface called energy directors. It is important to note that there are many variations of joint designs, which are detailed in Chapter 8. The cyclical deformation results in intermolecular friction and heating of the energy director, which eventually melts and flows to bond the parts.

Advantages:

- Fast cycle time (usually, welding time is less than 5 s.)
- Easy and accurate in process control (modern ultrasonic welders offer options for Statistical Process Control (SPC) as well as online diagnostics)
- Easy for automation
- Easy of re-tooling
- Precise welding of smaller parts is possible
- Using multiple welding heads, this process can cover various sizes of parts
- Able to join with a wide range of applications, such as inserts, staking, swaging and in addition to welding

Disadvantages:

- Very sensitive to process variations
- Sensitive to material characteristics such as moisture content and material grade etc.,
- Sensitive to joint design
- Far-field welding (see Chapter 8 for details) can be difficult for semi-crystalline materials
- Hermetical sealing of larger parts (bigger than one ultrasonic horn can handle) can be difficult

14.2.6   Vibration and Orbital Welding

Vibration and orbital welding involves clamping two parts together and holding one part stationary while the second part is vibrated in a linear or orbital motion. This motion initially results in slide friction heating of the faying surfaces until a molten layer is formed.   Once the faying surfaces are molten, viscous heating occurs resulting in an increase of the melt layer thickness. Pressure is maintained after the vibration motion is stopped resulting in cooling and solidification of the molten layer and welding of the parts. While vibration has been around for many years, orbital welding is becoming popular. Orbital welding has the advantage that unsupported internal walls can be joined and the cycle time is typically less compared to linear vibration welding.

Advantages:

- Fast process (3 to 15 s)
- Easily automated
- Able to weld medium/large parts (+500 mm)

- Robust process to overcome paint over-spray or other contaminants at the weld joint
- Relatively indepenedet of material properties

Disadvantages:

- Limited to flat joint areas (less 10° angle along the vibration direction is recommended)
- The application/part design must allow vibrational movement
- Difficult to weld thin, unsupported walls
- Weld bead can create sink mark
- It is difficult to hold thin and flat components during vibration
- Weld joint needs a solid foundation to maintain rigidity
- Selected materials produce particulates
- Appearance requirements can make special consideration for weld flash containment necessary

14.2.7   Spin Welding

In spin welding, one part is held stationary while the second rotating part is pressed against it. Initially, surface friction results in heating and melting of the joint surfaces. Once a molten layer is formed, viscous heating results in further melting and an increase of melt layer thickness. When the rotation is stopped the joint area cools and a weld is formed. While the most common mode of spin welding involves rotating the parts in a complete circle (360°), it is also possible to partially rotate the moving component through a pre-selected arc (<360°). This mode of spin welding is often called angular friction welding and allows a broader range of applications to be welded with spin welding.

Advantages:

- Fast process (2 to 5 s)
- Economical, relatively low capital costs
- Hermetic seal is possible
- Applicable to a wide range of materials

Disadvantages:

- Limited to part/joint size (small to medium size)
- Limited to circular geometries to achieve hermetical sealing
- For structural strength of the welds, joint design has to be determined carefully

14.2.8   Dielectric Welding

In dielectric or RF welding (radio frequency welding), an alternating electric field is applied to a lossy dielectric polymer, such as PVC or polyurethane. The polymer molecules vibrate and rotate in response to the electric field, resulting in intermolecular frictional heating. This results in melting of the polymer, which is then cooled under pressure to form a bond or seal.

Advantages:

*   Fast (5–10 s)
*   Economical
*   Easily automated
*   Ideally suited to thin and flat parts
*   Multiple layers can be welded simultaneously

Disadvantages:

*   Limited joint thickness
*   Limited joint complexity
*   Material limitations

14.2.9   Laser and IR Welding

In laser welding, different laser sources such as diode, YAG, and CO_2 can be used to heat the joint interface. In IR welding lamps and heaters can be used to heat the joint interface. This results in the formation of a molten layer at the interface. Pressure is applied to squeeze the molten layer and promote intimate contact while the polymer cools and solidifies. In addition, it is often possible cut the material in order to achieve a "cut and seal" operation.

Advantages:

*   Fast (1–10 s)
*   Non-contact heating
*   Simple joint design
*   Small heat affected zone
*   Through transmission welding (TTIr) technique can offer some unique benefits, such as no particulate generation, limited movement of the parts during welding and no part marking.

Disadvantages:

- Relatively expensive capital equipment
- Relative high maintenance
- Exhaust systems maybe needed (especially in higher power density, such as in cutting applications)
- Optical properties of materials is very critical

14.2.10   Microwave Welding

In microwave welding, a susceptor (e.g., composite of the polymer and either a conductive polymer, dipolar polymer, or ferromagnetic material) is placed at the interface and is heated by absorption of microwave radiation. The polymer near the susceptor melts and then pressure is applied squeezing the gasket out of the joint area. This results in molten polymer making intimate contact to form a joint.

Advantages:

- Relatively fast (5 to 20 s)
- High energy efficiency
- Non-contact heating
- Able to weld internal walls
- Complex joints are possible
- Amendable to unsupported internal walls
- Disassembly for recycling is possible.

Disadvantages:

- Shielding from the welder is required
- Uneven heating (mixing the microwaves and rotating sample can help)
- Not a fully developed process
- Consumable material is often required

Because this process is relatively new, there are only confidential applications that are being assembled with this technique. Thus, it is not possible at this time to provide an application example for microwave welding.

14.3   Process Selection

In addition to the characteristics of each welding process influencing the decision on selecting a process for an application, the process requirements are equally important to consider:

- Material characteristics
- Part size
- Joint geometry
- Manufacturing and production requirements
- Cost analysis

The following sections review each of these factors in more detail and help provide insight into how to select the best-suited process for a certain application.

14.3.1   Material Characteristics

As discussed in Chapter 1, thermoplastics are typically grouped into two categories based on their microstructure:

- Semicrystalline materials
- Amorphous materials

Because of their inherent differences, each group is better suited for different welding processes. For example, semicrystalline plastics typically require more energy to melt compared to amorphous thermoplastics. High-density polyethylene (HDPE) is a typical semicrystalline thermoplastic requiring a relatively high amount of heat to promote melting (because of the latent heat of fusion). Because of this requirement, proper heat control needs to be maintained in order to prevent polymer degradation, which could result in poor weld strength.

In contrast, polycarbonate (PC) is a typical amorphous plastic requiring less energy, but a higher welding temperature. In addition, the melt has a relatively high viscosity when the temperature is properly controlled, making wetting under pressure relatively easy. Thus, crystalline materials require a higher degree of energy control and amorphous materials, such as PC, require a higher degree of temperature control.

However, the melt flow behavior of these two plastics is not the only determining factor for selecting a process. In addition, their relative stiffness at room temperature can affect their weldability. Generally, semicrystalline materials are less stiff compared to amorphous materials. This makes semicrystalline materials less amendable to ultrasonic welding, especially in a far field mode. In addition, they tend to deflect more in vibration welding.

Additives such as fillers, pigments, flame retardants, mold release agents, etc. can greatly affect the weld strength. Small amount of fillers (less than 10%) can increase a material's stiffness and enhance its weldability for ultrasonic and vibration welding. However, higher levels of a filler material can increase melt viscosity that can interfere with the material flow and mix. When a radiant heating process is used, carbon black content can greatly affect the welding quality.

It is also common for fillers to collect or migrate to the bondline, which can compound their effects on a material's weldability. It is usually possible to detect such issues by evaluating a micro-cross section of the weld. In addition, when welding polymer blends (alloys), such PC/ABS, it is common that one of the materials, usually the continuous phase, will migrate to the bondline. This usually promotes the weld performance to more closely match one of the alloy components and not the bulk material. Again, a microstructure analysis of the bondline can often reveal such issues.

14.3.2   Part Size and Joint Geometry

Most direct heating processes such as vibration welding, hot plate welding, hot gas welding, and ultrasonic welding with multiple horns are suitable for both bigger and smaller size parts while non-contact and indirect heating processes such as RF welding, induction welding, IR welding, laser welding, and microwave welding are suitable for smaller and medium size parts. It is mainly because uniform heating of larger parts with widely distributed weld joint is very difficult to achieve with indirect heating processes.

For general reference, Table 14.1 shows a ranking of part size capacity for the various processes.

Table 14.1 Process Ranking of Size Capacity

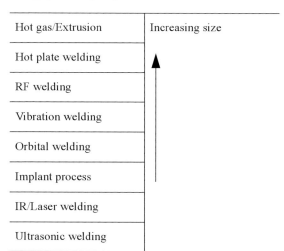

Hot gas/Extrusion	Increasing size
Hot plate welding	
RF welding	
Vibration welding	
Orbital welding	
Implant process	
IR/Laser welding	
Ultrasonic welding	

In addition to size, part geometry has a significant impact on process selection. For example, only parts with round faying surfaces can be assembled with spin welding. Also, far field weld joint of semicrystalline plastics must have relative planner or flat joints. Lastly, with vibration welding, angled surface and weld joints along the vibration movement should be minimized (max. < 10°).

14.3.3   Manufacturing Requirements and Cost Analysis

After all technical factors have been considered, the decisive factor driving the selection process is costs. There are five major costs components:

*   Initial capital costs

*   Maintenance costs

*   Consumables

*   Operation (electricity, air, and operator)

*   Reject costs

In order to determine how many welding system are needed, a balance must be found between process robustness, cycle time, and capital costs. While the cycle time of the process can often determine how many welding machines, operators, utility consumption, and equipment floor space available is needed, it is critical to look at the entire cost of the manufacturing process for the product life expectancy. For example, often it is more cost effective to invest in a higher cost process, in order to minimize the cost associated with rejects.

Repeatability of the process has a great impact on the scrap rate, which is an important measurable requirement in many plants. Repeatability of the process is very much dependent on component fixturing, design, and robustness in association with part size and weld joint geometry.

14.3.4   How to Select the Most Appropriate Welding Process?

*   Review characteristics (advantages and disadvantages) of all possible processes

*   Acquire and review product related information such as material characteristics of the component, part size, geometry, manufacturing requirements, and cost analysis

*   Compare the product related information and characteristics of the processes

Table 14.2 provides a general guideline to process selection. By applying all given criteria of the chart to a given application, the process maximizing the total points "scored" is best suited. Because this chart makes many assumptions and does not account for all possible factors, it should not be considered an absolute guide, therefore, the top scoring 3 to 5 processes should be considered.

Table 14.2 *General Points Addition for Various Processes and Application*

Part size (Linear weld length)	Ultrasonic welding	Laser/IR welding	Vibration welding	Orbital welding	Spin welding	Implant Welding (resistance & inductive)	RF Welding	Hot Plate welding	Hot Gas & Extrusion welding
2 to 20 cm	10	10	10	10	10	10	10	10	8
20 to 50 cm	7	8	10	10	8	10	10	10	10
50 to 100 cm	2	2	10	5	2	5	8	10	10
+100 cm	0	0	10	0	0	0	3	8	10

Required cycle time (The maximum values are centered on a typical cycle times for a given process)

−1 s	10	8	4	4	4	0	2	0	0
1 to 2 s	10	10	6	6	6	0	8	2	0
2 to 5 s	5	10	8	10	10	4	10	8	2
5 to 10 s	5	5	10	10	8	8	8	10	4
+ 10 s	2	5	8	8	6	10	2	10	10

Material

Amorphous	10	10	10	10	10	10	8	10	8
Crystalline	5	8	8	8	8	10	0	8	10

Part complexity

Simple	10	10	10	10	10	7	8	10	10
Multiple curvature	7	6	5	6	0	8	8	10	8
Very complex	5	5	3	5	0	8	2	8	8
Varies part to part	0	0	0	0	0	0	0	0	10

Part size (Linear weld length)	Ultrasonic welding	Laser/IR welding	Vibration welding	Orbital welding	Spin welding	Implant Welding (resistance & inductive)	RF Welding	Hot Plate welding	Hot Gas & Extrusion welding
Manufacturing requirements									
10–100/day	0	0	0	0	0	2	2	0	10
100–1000/day	2	8	2	2	2	10	5	2	8
1000–100000/day	8	8	7	7	7	8	8	7	5
+10000/day	10	5	5	5	5	5	5	5	0

14.4   Industrial Applications

In the following section, selected industrial applications are reviewed and details are provided why a particular joining process was selected for manufacturing.

14.4.1   Hot Plate/Heated Tool Welding

Figure 14.1 shows a typical automotive taillight that can be assembled using many of the processes detailed in this book. For example, it would be possible to use either adhesive bonding, vibration welding, or hot plate welding. However, adhesive bonding is relatively expensive and requires curing causing a longer cycle time. Using the general guideline table (Table 14.3), it is possible to determine the best process. For example, by appointing a value (x) for the relative application attribute, size, cycle time, etc., it is possible to then mark the process scores and add the points, see Table 14.3.

In this example, the points scored for each process are detailed in Table 14.4. It can be seen that vibration welding and hot plate welding are two possible methods with similarly high scores.

Table 14.3 Using Guideline Table for Process Selection for Taillight Application

Part size (Linear weld length)	Ultrasonic welding	Laser/IR welding	Vibration welding	Orbital welding	Spin welding	Implant Welding (resistance &)	RF Welding	Hot Plate welding	Hot Gas & Extrusion welding
2 to 20 cm	10	10	10	10	10	10	10	10	8
20 to 50 cm	7	8	10	10	8	10	10	10	10
50 to 100 cm	2	2	10	5	2	5	8	10	10
+100 cm (X)	0	0	10	0	0	0	3	8	10

Required cycle time (The maximum values are centered on a typical cycle times for a given process)

−1 s	10	8	4	4	4	0	2	0	0
1 to 2 s	10	10	6	6	6	0	8	2	0
2 to 5 s	5	10	8	10	10	4	10	8	2
5 to 10 s (X)	5	5	10	10	8	8	8	10	4
+ 10 s	2	5	8	8	6	10	2	10	10

Material

Amorphous (X)	10	10	10	10	10	10	8	10	8
Crystalline	5	8	8	8	8	10	0	8	10

Part complexity

Simple	10	10	10	10	10	7	8	10	10
Multiple curvature (X)	7	6	5	6	0	8	8	10	8
Very complex	5	5	3	5	0	8	2	8	8
Varies part to part	0	0	0	0	0	0	0	0	10

Part size (Linear weld length)	Ultrasonic welding	Laser/IR welding	Vibration welding	Orbital welding	Spin welding	Implant Welding (resistance &)	RF Welding	Hot Plate welding	Hot Gas & Extrusion welding
Manufacturing requirements									
10–100/day	0	0	0	0	0	2	2	0	10
100–1000/day	2	8	2	2	2	10	5	2	8
1000–100000/day (X)	8	8	7	7	7	8	8	7	5
+10000/day	10	5	5	5	5	5	5	5	0

Table 14.4 *Points Scored for Each Process for Taillight Application*

Process	Points
Ultrasonic welding	30
Laser welding	29
Vibration welding	**42**
Orbital welding	33
Spin welding	25
Implant welding	34
RF welding	35
Hot plate welding	**45**
Hot gas/extrusion welding	35

Vibration welding may cause part marking, promote high residual stress, and produce particulates. For these reasons, hot plate welding was selected as the process of choice. While it is relatively slow compared to vibration welding, the process can tolerate part-to-part variation caused by multi-cavity molding. In addition, the process can produce hermetic seals with no part marking and no particulate generation.

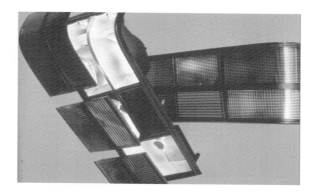

Figure 14.1 *Example of automotive taillights assembled with hot plate welding (Courtesy Branson Ultrasonics Corp.)*

14.4.2   Hot Gas and Extrusion Welding

Figure 14.2 shows a large chemical scrubber that is assembled using extrusion welding. The scrubber has to be resistant to acids and many other chemicals and therefore is fabricated from a crystalline material. In addition, it is critical that the welds be hermetic. By using Table 14.3, it is possible to determine which process is best-suited to assemble this application, see Table 14.5.

Figure 14.2 *Large chemical scrubber that is assembled with extrusion welding (Courtesy Wegener USA.)*

Table 14.5 Points Scored for Each Process for Scrubber Application

Process	Points
Ultrasonic welding	12
Laser welding	18
Vibration welding	26
Orbital welding	16
Spin welding	14
Implant welding	30
RF welding	9
Hot plate welding	34
Hot gas/extrusion welding	**48**

Because of the size of the application, hot gas or extrusion welding are the only processes suitable for this application.

14.4.3   Implant Induction Welding

Figure 14.3 shows a typical application assembled with induction welding, with Table 14.6 displaying the result of the point scoring technique.

Figure 14.3 Coffee pot assembled with induction welding (Courtesy Ashland Specialty Chemical Company Emabond Systems)

Table 14.6 *Points Scored for Each Process for Coffee Pot Application*

Process	Points
Ultrasonic welding	29
Laser welding	36
Vibration welding	35
Orbital welding	37
Spin welding	28
Implant welding	**46**
RF welding	33
Hot plate welding	**40**
Hot gas/extrusion welding	38

It is seen that implant welding and hot plate welding can both be used to assemble this application. However, induction welding was selected for this application because the joint was located internally. In addition, the final machine design allowed several part designs (models) to be welded on a single machine.

14.4.4   Ultrasonic Welding

Figure 14.4 shows a typical application joined with ultrasonic insertion. This method was selected because it allowed for a fast cycle time (2–4 s) without the need for designing a complex mold to accommodate inserts (insert molding).

Figure 14.5 shows a smaller application for ultrasonic welding, which normally requires high precision process control to avoid part damage and quick cycle times (less than 20 s including loading and unloading of the part). For this part, it was also important to control weld flash in order to meet the requirements for part appearance.

As seen in Table 14.7, both ultrasonic welding and laser welding could be used to assemble this application. However, because of cost considerations, ultrasonic welding was selected. It is also important to note that while orbital welding and spin welding both scored relatively high, neither one is applicable because of the part geometry, which does not allow for relative motion between the two components.

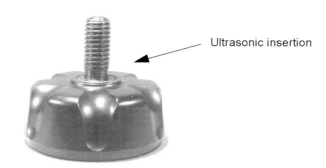

Ultrasonic insertion

Figure 14.4 *Example of ultrasonic insertion*

Figure 14.5 *Example of ultrasonic welding of a small application that requires a hermetic seal (make-up Application)*

Table 14.7 *Points Scored for Each Process for Make-up Application*

Process	Points
Ultrasonic welding	**50**
Laser welding	**45**
Vibration welding	41
Orbital welding	43
Spin welding	43
Implant welding	32
RF welding	39
Hot plate welding	37
Hot gas/extrusion welding	26

14.4.5   Vibration and Orbital Welding

Figure 14.6 shows an in-dash air duct for the automotive industry assembled with vibration welding. In this application, a hermetic seal around the weld joint area was required with a weld joint continuous across the part perimeter. Based on the result of the point scoring technique, see Table 14.8, vibration and hot plate welding could be identified as the best-suited methods to assemble this application.

Table 14.8 Points Scored for Each Process for In-Dash Application

Process	Points
Ultrasonic welding	30
Laser welding	33
Vibration welding	**45**
Orbital welding	40
Spin welding	35
Implant welding	38
RF welding	32
Hot plate welding	**45**
Hot gas/extrusion welding	39

The welding cycle time had to be less than 20 s, which is not easy to achieve with the hot plate welding process. Thus, only vibration welding was able to assemble the part quickly and produce a hermetic seal.

Figure 14.6 *In-dash air duct assembled with vibration welding (Courtesy Branson Ultrasonics Corp.)*

Figure 14.7 shows an automotive intake manifold, which requires a continuous hermetic seal around the weld joint areas. In previous designs for this application lost-core molding was used, but because of cost considerations molding the manifold in two halves and welding the halves together was considered. Initially there were problems with welding such a large part requiring high strength and a hermetic seal. Using the point scoring technique to determine which process is most applicable, led to the conclusion that both vibration welding and hot plate welding can be used (similar to Table 14.8). In practice, vibration welding has become a popular method for assembling this application, because the process is quicker compared to hot plate welding and the weld flash (which can affect air flow) can be controlled by using proper joint design. In addition, highly contoured weld surfaces in three dimensions make uniform heating difficult when using the hot plate welding process.

Figure 14.7 Intake manifold assembled with vibration welding (Courtesy Branson Ultrasonics Corp.)

Figure 14.8 shows an under-the-hood electronic housing that is assembled with orbital welding. The process allows un-supported internal walls to be welded quickly, with a cycle time of less than a few seconds. Linear vibration welding would not facilitate the welding of internal welds and hot plate welding would be too slow to meet manufacturing requirements.

Figure 14.8 Automotive electronic housing assembled with orbital welding

14.4.6   Spin Welding

Figure 14.9 shows a circular weld joint in a hose application. As seen in Table 14.9, laser welding, orbital welding, and spin welding can be used to assemble the application. However, with the current joint design, an interference joint design prevents relative motion needed for orbital welding. On the other hand, the weld geometry makes this application well suited for the spin welding process. Laser welding is not the process of choice because the process benefits, such as reduced generation of particulates, would not justify the added equipment costs. It is important to note that the part geometry is not circular and the tooling had to be designed to accommodate the geometry and offset the weight distribution to assure the center of mass was located on the center of rotation. In addition, the cycle time for this application was less than 20 seconds including loading and unloading of the parts.

Table 14.9 Points Scored for Each Process for Hose Assembly Application

Process	Points
Ultrasonic welding	38
Laser welding	**46**
Vibration welding	43
Orbital welding	**45**
Spin welding	**45**
Implant welding	39
RF welding	36
Hot plate welding	43
Hot gas/extrusion welding	35

Figure 14.10 shows a photograph of a small tank assembled with spin welding. Again, the process was quick (less than 3 s welding time) and it allowed press-fit of welded parts, which helped locating the parts prior to welding. It should be noted that although circular parts can also welded by the ultrasonic welding process, this application had a lid with a contoured surface, which made it difficult to be welded by the ultrasonic welding process. Also, the material (a crystalline material) was not well suited for ultrasonic welding.

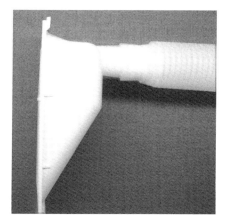

Figure 14.9 Hose assembly welded with spin welding

Figure 14.10 Small tank assembled with spin welding

14.4.7   RF Welding

Probably the most common applications of RF welding for assembly are medical bags, see Figure 14.11.

As seen in Table 14.10, hot plate welding and RF welding are two techniques that would be amenable for assembly of these applications. It should be noted that in this application, impulse welding, a special type of hot plate welding would have to be used. In impulse welding, the heat source is applied to the exterior of the weld sample by a small wire coated with a non-stick material, such as Teflon. The wire is quickly heated by resistance heating and the heat melts the bulk of the material in order to make a weld. However, industry primarily uses RF welding for sealing medical bags. The main advantage of the process is that hermetic seals are easily achieved on both small and large products. In addition, the weld flash is smooth (to reduce turbulent flow within the bags) and fully attached, so that there are no particulates to break free that could contaminate the fluid. Lastly, because many of the bags have similar shapes and sizes, it is easy for the manufacturer to re-configure the machine to weld various products. Because of these advantages and the long history of RF welding in this industry, PVC has been nearly the exclusive material used to fabricate these products.

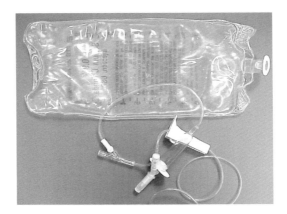

Figure 14.11 Medical bags and IV's sealed with RF welding

Table 14.10 Points Scored for Each Process for Hose Assembly Application

Process	Points
Ultrasonic welding	32
Laser welding	36
Vibration welding	40
Orbital welding	38
Spin welding	33
Implant welding	35
RF welding	**42**
Hot plate welding	**45**
Hot gas/extrusion welding	33

14.4.8   Laser and IR Welding

Figure 14.12 shows a typical application welded with TTIr, an electronic consumer product. The clear window is welded to a black housing. Based on the point scaling technique, there are five possible joining techniques for this application: ultrasonic welding, laser welding, vibration welding, orbital welding, and hot plate welding.

Figure 14.12 Consumer electronics product welded with TTIr

TTIr (laser welding) was selected because it allowed the part to be hermetically sealed without the generation of particulates or flash that would be visible to the consumer through the window. While the other possible techniques may have proved feasible, they come with a higher risk for excessive flash or part marking.

One the first applications industrially assembled using IR welding was a moisture filter, see Fig. 14.13. In this application, a small membrane was "heat sealed (impulse welding)" to a frame. This assembly was then welded to an outer housing. Initial designs evaluated the use of ultrasonic welding; however, there were issues with filter damage because of the vibrations. Laser welding was evaluated and proved a viable method. The process (TTIr) was fast (1–2 s) and did not damage the relatively delicate membrane.

Figure 14.13 Delicate filter assembly welded with laser technique

Table 14.11 *Points Scored for Each Process for Consumer Electronic Application*

Process	Points
Ultrasonic welding	**43**
Laser welding	**43**
Vibration welding	**47**
Orbital welding	**45**
Spin welding	42
Implant welding	42
RF welding	42
Hot plate welding	**47**
Hot gas/extrusion welding	35

15   Testing of Weld Joints

Christian Bonten and Carsten Tüchert

The quality of weld joints can be evaluated by different methods. In order to apply the appropriate method for any given welded application it is necessary to determine the main loading and design requirements. Whether the loading will require a high strain-behavior, short-term or long-term strength, leak-tightness, or optical criteria such as cosmetic appearance determines the test method of choice.

Overall, testing can be divided into non-destructive and destructive testing. Table 15.1 gives an overview of the different testing methods.

Generally, an exact determination of a weld's quality is only possible by destructive testing. In addition, different test methods are usually required to determine the weld strength or the weld stiffness. For example, in order to determine weld strength, a tensile test is usually performed. While this test does provide information on stress and strain relations, it is usually recommended to use a bend test to determine a weld's stiffness. In addition, none of these tests provides information on a weld's ability to withstand cyclic loading, which can only be evaluated by a fatigue test. In addition, only creep tests provide information on the long-term behavior of weld seams. However, weld performance can only be reliably determined after the weld is tested in the application under representative environmental conditions.

Despite the efforts of many investigators, currently there are no non-destructive tests that allow the prediction of long-term weld performance.

In general, destructive test methods can be divided into standardized and non-standardized testing methods. Standardized methods are used mostly for semi-finished products (pipes, sheets, etc.), because here it is possible to obtain specimens that comply with a given standard. Standardized methods have the advantage of comparability, e.g., with a base material sample. Because of inherent differences between samples, part complexity, and part geometry, most mass-produced products require special testing methods. Therefore, the test method of choice is tailored to each single product. Thus, the results of these tests are only applicable to a particular application and a comparison of weld quality to other applications is usually not possible. Despite these limitations, non-standardized tests allow optimizing the welding parameters based on the weld quality test. Usually these special test procedures and their loading conditions are designed to closely represent the loading conditions on the weld once placed in the application.

Table 15.1 Overview of the Different Testing Methods

Category	Method	Advantage	Disadvantage
Non-destructive testing	– Visual testing (according to DVS-2202-1 [1])	– Non-destructive method – Fast detection of imperfections on the surface of the weld joint	– Requires high level of experience and skill – No evaluation of weld seam quality (short- or long-term)
	– X-ray testing – Ultrasonic testing – Leak-tightness testing – High-voltage testing	– Non-destructive method – Fast detection of imperfections inside the weld joint	– No evaluation of weld seam quality (short- or long-term)
Destructive testing	– Tensile test – Impact test – Bend test – Creep test	– Quantitative information of weld stress strain behavior (e.g. tensile, impact, bend.) – Quantitative information of weld strength (short- or long-term)	– Requires the destruction of the weld seam – Large preparation effort maybe needed

15.1   Non-Destructive Testing

The following test methods are used for non-destructive testing:

- Visual testing,
- X-ray testing,
- Ultrasonic testing,
- Leak-tightness testing,
- High-voltage testing.

The main advantage of non-destructive testing lies in the fact that the sample remains intact and can be used after inspection. However, the main limitation of these tests is that they provide limited information in terms of weld strength. Typically, these test methods are used to determine if imperfections or inhomogeneous material areas are present inside the weld area [2]. Each of the above mentioned techniques is reviewed in detail in the following sections.

15.1.1   Visual Inspection

While visual testing/inspection is the least expensive testing method, it can only be used to detect flaws on the surface of the weld seam or on the faying surfaces during the welding process. Visual inspection can be categorized in two groups:

- *Active weld inspection*, which means for the operator to inspect the joint area during the welding process, and
- *Post-weld inspection*, which involves inspection of the weld after it has been completed.

With both categories (active and post-weld inspection), only certain defects affecting weld quality can be detected. For example with active weld inspection, discolorations are often an indication for thermal degradation. Deviation of the weld shape (angular misalignment, bead camber, waviness, etc.), bubbles, notches, scratches, etc. can also be correlated to weld defects [1] affecting weld strength. However, it is important to note that only an experienced operator will be able to reliably detect weld flaws during the welding process, which poses another limitation of visual active weld inspection.

Post-weld inspection allows the re-crystallization and solidification effects to be evaluated, which can change the appearance of the weld [2]. Table 15.2, 15.3 and 15.4 show examples of imperfections for welding processes for different semi-finished products. The reader is directed to DVS 2202-1 for more details. This standard applies for high-density polyethylene (HDPE), polypropylene (PP), polyvinylchloride (PVC), and polyvinylidene fluoride (PVDF). For these materials, the imperfections are classified in three groups: Group I is used for applications with high requirements on quality, Group II represents medium requirements, and Group III low quality and/or load baring requirements. It is important to note that the definition of the groups was made for simplification reasons only and that the application and the performance requirements determine the actual acceptance criteria. When selecting a group, it is important to consider the following factors [1]:

- The type of stress (static or dynamic)
- The material (viscous or brittle)
- The operating conditions (stationary or moving)
- The production environment (workshop, construction site, or welding in constrained areas)
- The potential danger (dangerous gases or fluids)

If more than one defect is detected (see Table 15.2 and 15.3) it is important to note that the total reduction in weld performance is often greater than the sum of the detrimental effects of each defect.

Table 15.2   *Evaluation Groups for Welding of Semi-Finished Products [1]*

External state of joint	Process	Description	Evaluation groups
Welding flash notches	Heated tool butt welding/hot plate welding	Continuous or local notches lengthwise to weld with notch root in base material, caused by, for example – insufficient joint pressure – warming-up time too short – cooling time too short	Not permissible
Notches and flutes	Heated tool butt welding/hot plate welding	Notches in edge of base material, lengthwise or crosswise to weld, caused by, for example: – clamping tools – incorrect transport – fault in edge preparation	Permissible Group I: $\Delta s \leq 0.1$ s, but max. 0.5 mm Group II: $\Delta s \leq 0.1$ s, but max. 1 mm Group III: $\Delta s \leq 0.15$ s, but max. 2 mm
Mismatch of joint faces	Heated tool butt welding/hot plate welding	Joint faces are displaced relative to one another or thickness variations are not adjusted	Permissible Group I: $e \leq 0.1$ s, but max. 2 s Group II: $e \leq 0.15$ s, but max. 4 s Group III: $e \leq 0.2$ s, but max. 5 s

External state of joint	Process	Description	Evaluation groups
Angular deflection	Heated tool butt welding/hot plate welding	For example: – machine fault – layout/fixturing fault	Permissible Group I: $e \leq 1$ mm Group II: $e \leq 2$ mm Group III: $e \leq 4$ mm
Narrow, excessive welding flash	Heated tool butt welding/hot plate welding	For example: – excessive and sharp edged weld bead over part or all of weld length or weld girth due to wrong welding parameters, especially caused by – excessive joint pressure (with polyolefins only)	Not permissible

Table 15.3   *Description of Evaluation Groups for Welding of Semi-Finished Products [1]*

External state of joint	Process	Description	Evaluation groups
Faulty welding bead formation	Heated tool socket welding	Variable welding bead formation or no welding bead present on one or both sides (over part or whole of weld length) due to – unprocessed joint faces – contaminated joint faces – heated tool temperature too high	Not permissible

Table 15.3   *Description of Evaluation Groups for Welding of Semi-Finished Products [1] (Continuation)*

External state of joint	Process	Description	Evaluation groups
Angular deflection (defect of form)	Heated tool socket welding	Pipe welded into fitting at angle on one or both sides, or with slightly faulty gripping, caused by, for example: – machine fault – layout fault	Permissible Group I: $e \leq 1$ mm Group II: $e \leq 2$ mm Group III: $e \leq 4$ mm
Weld too high	Hot gas welding	Exceeds standard fillet weld thickness of: $a = 0.7\,s$	Permissible Group I: $b \leq 0.4\,a$, but max. 6 mm Group II: $b \leq 0.5\,a$, but max. 9 mm Group III: $b \leq 0.6\,a$, but max. 12 mm
Weld too low	Hot gas welding	Exceeds standard fillet weld thickness of: $a = 0.7\,s$	Group I: not permissible Group II: permissible if nominal dimension a is slightly undershot locally $b \leq 0.15\,a$ Group III: permissible if nominal dimension a is undershot locally $b \leq 0.3\,a$

Table 15.4   *Description of Evaluation Groups for Welding of Semi-Finished Products [1]*

External state of joint	Process	Description	Evaluation groups
Cracks	Hot gas or extrusion welding	Isolated cracks or groups of cracks with or without branching, running lengthwise or crosswise to weld They may be located: – in the weld – in the base material – in the heat affected zone	Not permissible
Welding flash notches	Hot gas or extrusion welding	Continuous or local flat deformation lengthwise to weld, caused by, for example: – faults in welding shoe – faults in welding unit guidance	Group I: not permissible Group II and Group III: locally permissible if $k > 0$
Edge notches	Hot gas or extrusion welding	Notches in base material along weld, caused by, for example: – penetration of weld shoe edges – processing of edge zone	Group I: locally permissible if ending flat and $\Delta s \leq 0.1\ s$, but max. 1 mm Group II: continuously permissible if ending flat and $\Delta s \leq 0.1\ s$, but max. 2 mm Group III: continuously permissible if ending flat and $\Delta s \leq 0.2\ s$, but max. 3 mm

Table 15.4   *Description of Evaluation Groups for Welding of Semi-Finished Products [1] (Continuation)*

External state of joint	Process	Description	Evaluation groups
Welding material overflow 	Hot gas or extrusion welding	Welding material overflow on one or both sides, usually without sufficient fusion to base material	Group I: not permissible Group II: locally permissible in small numbers if $\Delta b \leq 5mm$ Group III: continuously permissible if $\Delta b \leq 5$ mm
Angle deflections (defects of form) 	Hot gas or extrusion welding	Length, L, of shorter side is decisive for permissible deflection	Group I: permissible up to $\Delta \leq \pm 1$ Group II: permissible up to $\Delta \leq \pm 2$ Group III: permissible up to $\Delta \leq \pm 3$

15.1.2   X-Ray Testing and Ultrasonic Testing

Imperfections such as crazes, voids, porosity, and solid inclusions can be detected by X-ray and ultrasonic testing. In order for the method to be effective, the imperfection's density must differ from the plastic's density significantly and the imperfection's size must be greater than a minimal detectable size. The minimal size is usually determined by test procedure, defect orientation, equipment, and defect type. It is important to note that X-ray and ultrasonic testing techniques cannot be used to optimize process parameters. In addition, defects related to microstructure, such as cold joints or excessive squeeze flow cannot be detected with standard X-ray or ultrasonic testing methods.

Because these test methods are relatively labor intensive and costly, they are typically used for applications with high safety and reliability requirements [2, 3]. X-ray testing is usually limited to pressure vessels and pipelines that carry hazardous materials. It is not commonly used for standard manufactured components or mass produced products.

X-ray testing of plastic components is similar to the inspection of other materials, such as metal weldments, except that the radiation intensity is usually lower because plastics typically have relatively low densities. The setup for X-ray testing is seen in Figure 15.1, where a radiation source, usually an X-ray tube, transmits radiation through the sample and to a sensor, typically an X-ray sensitive film, or an X-ray sensitive camera capturing an image. Any defects with a different density will produce an image.

Ultrasonic testing involves transmitting high frequency sound waves (1 to 100 MHz) through the weld, relying on refraction and reflection of the sound wave to detect defects displaying different densities. There are two basic set-ups for ultrasonic testing:

• A transmitter and a receiver placed relative to each other so that defects will change the intensity and travel time of the sound wave (time of flight (TOF), see Fig. 15.2). In this configuration, it is common for the transmitter and receiver to be placed opposite each other with the weld between them. If it is not possible to make intimate contact between the transmitter/receiver and the sample, the test can be completed in a water bath to reduce attenuation of the sound wave.

• An ultrasonic transducer, called a transceiver, which both emits and receives the sound waves. This set-up is useful when access to the weld is limited to one side (see Fig. 15.2).

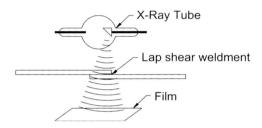

Figure 15.1 *Typical layout for X-ray testing of plastic welds*

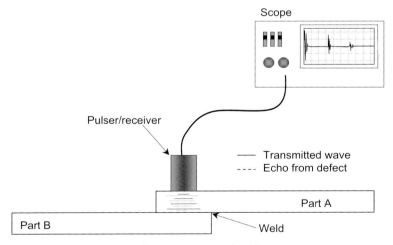

Figure 15.2 *Experimental set-up for ultrasonic testing of welds*

Table 15.5 shows whether ultrasonic and X-ray test procedures are suited for plastic materials typically used for applications with high safety and reliability requirements:

Table 15.5 *Suitability of Ultrasonic and X-Ray Testing for Selected Plastic Materials*

	Ultrasonic testing	X-ray testing
HDPE	Good	Acceptable
PP	Bad	Acceptable
PVC	Often acceptable	Acceptable

It is not common for X-ray and ultrasonic testing to be used with other materials. Additional information can be found in DVS 2206 [3].

15.1.3   Leak-Tightness Testing

There are many variations of leak-tightness testing depending on the medium used to pressurize the part, such as water, air, or other gases. This test method is mostly used for welded pipes, chemical containers, and vessels, because in these applications, small leaks can result in enormous potential of danger. Table 15.6 show the different leak-tightness testing methods. In addition, it is common to complete the test with a negative pressure, or vacuum. One of the simplest leak-tightness tests involves placing a hermetically seal application into hot water and watching the weld joint for bubbles. The hot water heats the container, which in turn heats the gas trapped in the application and causes a temporary increase of internal pressure resulting in bubbles.

Table 15.6 *Leak-Tightness Testing Methods in Accordance to DVS [3]*

Application	Pressure	Medium	Preparation and testing
Un-pressurized parts containing fluids	Ambient pressure	Water	– all openings, except the filling hole, are closed – the part is filled with water – the filling hole is kept open – the water level is marked – all weld seams and closed openings are controlled for leaks – leaks must be marked and repaired – if no leaks are detected, the second testing is performed 24 hours after the first testing – recognizable leaks must be repaired

Application	Pressure	Medium	Preparation and testing

Parts, subjected to overpressure	Overpressure	Water	– the filling hole is prepared with an inlet valve, manometer and a bleeder/vent
			
			– the part is filled with a pump, testing pressure $p_t = 1.3 \times$ process pressure
			
			  pipes are filled with $p_t = 1.5 \times$ process pressure
			
			  Note: only water or another non-compressible fluid should be used to minimize the danger of an explosion
			
			– the inlet valve is closed after reaching p_t; p_t is kept up for 1 hour
			
			– decompression in the first minutes without any leaks is possible, because of elastic extent
			
			– after the first decompression, the pressure must built up again to p_t
			
			– leaks are proven to exist, when the pressure falls again after pressure increase
			
			– recognizable leaks must be repaired

Parts, subjected to overpressure Ventilation pipes	Overpressure	Air	– in order to minimize the danger of explosion always use blow-up valves when testing with over-pressure – the same preparation as for over-pressure testing with water, $p_t =$ 100 mm water column over atmospheric pressure, for venti-lation pipes $p_t =$ process pressure – small parts or hollow bodies are tested in a water bath, leaks can be detected through bubble generation

Table 15.6 *Leak-Tightness Testing Methods in Accordance to DVS [3] (Continuation)*

Application	Pressure	Medium	Preparation and testing
			– when testing large parts, the weld seam and other joints are sprayed with watery soap solution (do not use surfactants, they may cause environmental stress cracking), leaks can be detected by bubble generation
Parts subjected to negative pressure	Negative pressure	Air	– after caulking of all openings, the pressure is decreased to p_t = 0,7 x process pressure (with vacuum pumps or water jet pump) – because of implosion danger, use only with suitable safety equipment – the increase of pressure should not exceed a previously specified value during the next hour.
Parts for high requirements	Overpressure	Halogen alkane	– in order to test for leak-tightness under high pressure the same procedure as for testing under overpressure with air can be performed, however the media used are halogen alkanated gases like FRIGEN 12® or a frigen-nitrogen mixture – this gases diffuse through micro-pores or crazes, leaks can be detected with suitable equipment – after testing, the plant must be completely rinsed with air – consider: the testing is only possible in a halogen-free atmos-phere; chlorinated hydrocar-bons, colorants, some softening agents or to hot welded PVC can extract halogens and thus falsify the result

15.1.4   Testing with High Voltage

Another variant of non-destructive testing is high voltage testing, commonly called a "spark test". This technique is a very practicable alternative to other leak-tightness tests. It is used for linings (pond liners and pipe liners), in where the welds are only accessible from one side (e.g., chemical protective coatings in glass-fiber reinforced plastic vessels) [2, 3].

In order to perform the spark test when the weld is only accessible from one side, the weld must be coated with an electrically conductive medium, such as metallic wires, conductive fibers, or metal foils. In the case of metal pipes or linings of metal vessels, the metal application itself will act as a counter electrode. When an arc can be made through the weld, this is an indication for the presence of a leak.

Testing with high voltage is usually difficult with polar thermoplastics, such as PVC. Polar thermoplastics can generate heat by high voltage similar to radio-frequency welding (see Chapter 11). In extreme cases, the heat can thermally degrade the weld. Table 15.7 shows recommended testing voltages for different wall-thicknesses.

Table 15.7 *Test Voltage for Different Wall-Thicknesses and Materials [3]*

Test voltage	Sparking distance in air (non-homogenous field)	Guidelines for testing with DC voltage		
		Glass	Natural rubber	Thermoplastics
[kV]	[mm]	[mm]	[mm]	[mm]
1	0.5	0.1	0.1	0.1
2	1	0.1	0.1	0.1
3	1.6	0.2	0.2	0.2
4	2.3	0.3	0.3	0.3
5	3.1	0.4	0.4	0.5
6	4	0.5	0.6	0.7
7	5	0.6	0.8	0.9
8	6.1	0.8	1.0	1.2
9	7.3	1.0	1.3	1.5
10	8.6	1.2	1.6	1.5

Table 15.7 *Test Voltage for Different Wall-Thicknesses and Materials [3] (Continuation)*

Test voltage	Sparking distance in air (non-homogenous field)	Guidelines for testing with DC voltage		
		Glass	Natural rubber	Thermoplastics
[kV]	[mm]	[mm]	[mm]	[mm]
11	10	1.5	1.9	2.3
12	11.3	1.7	2.12	2.7
14	14		3	3
18	19		4	4

15.2   Destructive Testing

Destroying the test sample is the only true way of determining weld quality, which in itself is the major disadvantage of these techniques. Destructive testing is also used to determine optimal welding parameters.

The standard destructive tests are tensile test, bend test, tensile impact test, and tensile creep test. With standardized specimens, destructive tests allow the examination of material and process parameters independently from the joint. These parameters can be used for production and joint design optimization. Standard tests, impact load tests, tests under internal pressure, bursting pressure tests, and tests performed in a medium can be directly related to the requirements of the application [1]. With mass-produced applications, standardized specimens and procedures are often not available because of complex part geometry, which do not allow an analytic calculation of the weld load generated during the testing.

Each destructive testing technique is detailed in the following sections.

15.2.1   Tensile Test

Tensile tests have the advantage to be quick, relatively simple, and to provide quantitative information on weld quality.

For the evaluation of mechanical properties, specimens of the weld and of the base material are both tested and the results are compared. The geometry shown in Fig. 15.3 is a common specimen used in tensile tests. In addition, it is possible to use the American Welding Society's (AWS) G1.2 standardized sample (American National Standard) [4]. This sample is commonly used with ultrasonic welding, however, it can also be used with other processes, such as vibration welding, laser welding, and hot plate welding [5].

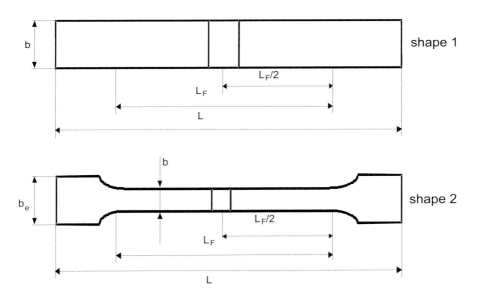

Figure 15.3 Tensile test specimen [6]

Figure 15.4, shows two of the common sample configurations, detailed in the AWS standard, an energy director. It is important to note that the standard also details a shear type joint design. The reader is referred to the standard for additional configuration and testing details.

Usually, the ultimate load sustained by the specimen is recorded. It is also possible to measure the strain induced in the sample by placing extensometers on the sample during testing. Table 15.8 shows typical dimensions for the test specimens. Usually the samples are tested without the weld beads (weld flash), because the flash can offset the results by carrying a portion of the applied load. If it is decided not to remove the weld flash, this should be reported with the weld data results. If the welded specimen does not fail in the weld seam, the specimen can be weakened in the joint by machining a hole in the weld area to act as a stress concentration point. Usually, a hole with a diameter of 3–4 mm is recommended [6, 7]. If this procedure is used, it is important to drill a similar hole in the base material sample to provide comparable test results.

The crosshead speed of the grips should be based so that yielding of the base material is reached after one minute of initiating the test. Table 15.9 provides recommended crosshead speeds for selected materials. For a statistical analysis, a minimum of 5 specimens of each weld seam and of the base material must be tested.

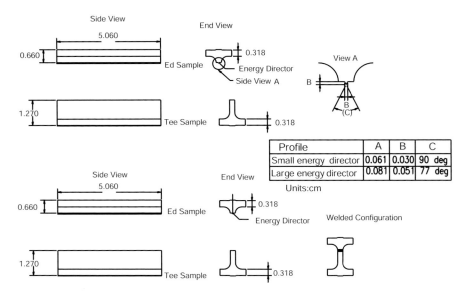

Figure 15.4 *Details of AWS test samplee*

In general, the tensile test is performed at a temperature of 23 ± 2 °C. Equation 15.1 can be used for the determination of the short-term tensile welding factor f_Z. The welding factor is the quotient of the force that can be supported by the weld seam and the force that can be supported by the material itself (without weld).

$$f_Z = \frac{F_W}{F_B}$$
(15.1)

where:
f_Z = short-term welding factor [-]
F_W = breaking force of the weld seam [N] (average value)
F_B = breaking force of the base material [N] (average value)

For other specimen geometries, the following formula is used:

$$f_Z = \frac{F_W}{b_W \cdot h_W} \cdot \frac{b_B \cdot h_B}{F_B} = \frac{F_W}{F_B} \cdot \frac{b_B}{b_W} \cdot \frac{h_B}{h_W}$$
(15.2)

h = thickness of the specimen [mm]
b = width of the specimen [mm]

Subscript W denotes weld and substricpt B denotes base material.

Table 15.8 *Dimensions of Tensile Test Specimens [6]*

Thickness h [mm]	Specimen 1 [mm]			Specimen 2 [mm]			
	B	L_f	L	B	L_f	L	b_e
≤ 10	15	120	≥ 170	10	115	≥ 170	20
> 10	30	120	≥ 300	30	115	≥ 300	40
> 20	1.5 x h	200	≥ 400	1.5 x h	200	≥ 400	80

Table 15.9 *Recommended Test Speed for Selected Thermoplastics [6]*

Material	Test speed [mm/min]
PE-HD	50
PP, PVDF	20
PVC	10

It is also important to describe and note the type of failure of the specimens after the test [2, 6]. Some of the items that should be noted include:

- Failure location
- Failure mode (ductile or brittle)
- Failure surface appearance

Tensile tests can provide information on short-term tensile strength of a weld, but they cannot predict long-term performance. It is important to note that long-term performance is often more important in order to determine weld performance.

In Table 15.10 the minimum short-term tensile welding factors f_Z for several materials and processes used to weld semi-finished products are listed.

As previously mentioned, it is possible to measure the strain during a tensile test by attaching extensometer to the sample. By integrating the stress times the strain during the tests, it is possible to calculate the energy required to break the sample. This energy is referred to as the "energy to failure". It has been shown that the energy to failure is more sensitive to fluctuations in welding parameters than is the tensile strength [9]. For example, welds made with varying welding parameters may have similar or identical tensile strength; however, they may have widely varying energy to failure values. Thus, by optimizing a welding process based on energy to failure rather than on tensile strength, the increased resolution (sensitivity) between weld quality and welding parameters allows better optimization of the welding process.

Table 15.10 *Minimum Requirements of the Short Term Tensile Welding Factor f_Z [8]*

Welding Process	Short term tensile welding factor f_Z [-]				
	PE-HD	PP	PVC-C	PVC-HI, PVC-U	PVDF
Heated tool welding	0.9	0.9	0.8	0.9	0.9
Extrusion welding	0.8	0.8	–	–	–
Hot-gas welding	0.8	0.8	0.7	0.8	0.8

15.2.2   Tensile Impact Test

The tensile impact test provides quick information regarding weld strength at high strain rates. In addition, if the strain is measured the energy required to break the weld can be calculated. The tensile impact welding factor is the ratio of the energy required to break the weld and the energy required to break the base material (without a weld). By examining the appearance of the fracture surface, it can be determined whether the failure is brittle or ductile.

Equation 15.3 can be used for the determination of the tensile impact welding factor f_{SZ}:

$$f_{SZ} = \frac{W_{ZW}}{W_{ZB}}$$

(15.3)

f_{SW} = tensile impact welding factor [-]
W_{ZW} = impact energy of the weld seam [J/cm²] (mean value of the tested specimens)
W_{ZB} = impact energy of the base material [J/cm²] (mean value of the tested specimens)

For deviating geometries of the specimens, Eq. 15.3 can be redefined as seen in Eq. 15.1:

$$f_{SZ} = \frac{W_{ZW}}{b_W \cdot h_W} \cdot \frac{b_B \cdot h_B}{W_{ZB}} = \frac{W_{ZW}}{W_{ZB}} \cdot \frac{b_B}{b_W} \cdot \frac{h_B}{h_W}$$

(15.4)

h = thickness of the specimen
b = width of the specimen

Figure 15.5 shows a typical specimen used for impact testing. The specimen should be at least 4 mm thick. Depending on the requirements of the application, the specimen can be tested with or without the weld flash. However, it is important to record the removal of flash in the test result data. If the welded specimens do not fail in the weld zone, the specimen can be locally weakened by machining a 3–4 mm diameter [10] hole through the sample at the weld interface. If this method is used to promote the proper failure mode, it is important that the base material is subjected to the same procedure. Because the test data are usually distributed relatively widely, a minimum of 10 specimens of each weld seam and of the base material should be tested [2, 10].

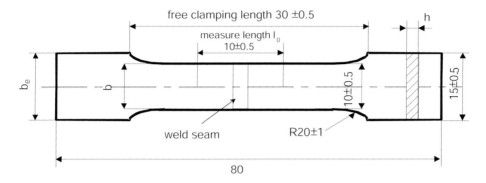

Figure 15.5 *Tensile impact specimen [10]*

Usually, the permanent fracture elongation of the sample is measured after the test. It can be measured by laying the broken parts together and measure the length. Equation 15.5 is used for the determination of the permanent fracture elongation in percent [10]:

$$\varepsilon_{bl} = \frac{l_{bl} - l_0}{l_0} \cdot 100\% \qquad (15.5)$$

l_0 = original measuring length [mm]
l_{bl} = measuring length after testing [mm]

15.2.3   Bend Test

Compared to many of the other short-term testing methods, bend tests provide a relatively higher level of confidence for the evaluation of the weld seam quality. During a bend test, the weld seam is loaded locally under tensile, compression, and shear resulting in different strains in the joint. This produces loads similar to the loading conditions generated in the final product. However, an exact measure of the weld quality is only possible by testing the welds in numerous loading conditions. The bend test gives a "high resolution" of the weld seam quality, but the results are only qualitative and only appropriate for comparative evaluations. Although the bend test can be used to optimize welding parameters, it does not provide characteristic values for the design of the joint or part [2, 11].

Figure 15.6 and Table 15.11 show the test arrangement and the design of the specimens for common bend tests. Figure 15.7 shows the cross-section of specimens from welded pipes and their dimensions. The specimens can be cut out of the pipe radially or parallel, relative to the axis of the pipe. The weld flash must be removed on the side where the test ram applies the load to the specimen. Also, the specimens must be prepared with chamfers that run trough the weld bead (Fig. 15.8).

The bend test is usually performed at a temperature of 23 ± 2 °C. For statistical reasons, a minimum of 6 specimens should be tested. For welded plates, 3 specimens can be tested for each side of the weld. For welded pipes, the inside is subjected to extension and should be tested in similar loading conditions. Table 15.12 shows the test speed/ram speed for selected thermoplastics.

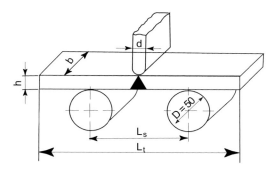

Figure 15.6 *Test arrangement for the bend test [11]*

Table 15.11 *Dimensions of the Test Arrangement and the Test Specimens [11]*

Test specimen				Distance between axes of roller L_S [mm]	Diameter of ram a [mm]
Thickness s [mm]	Width b [mm]		Minimum length L_1 [mm]		
Nominal value	Pipe	Plate			
$3 < s \leq 5$	0.1 x d [(1)] min.: 6 max.: 30	20	150	80	4
$5 < s \leq 10$		20	200	90	8
$10 < s \leq 15$		20	200	100	12.5
$15 < s \leq 20$		30	250	120	16
$20 < s \leq 30$		30	300	160	25

(1) Nominal diameter

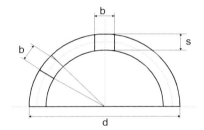

Figure 15.7 Cross sections of specimens cut from pipes [11]

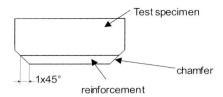

Figure 15.8 Geometry of the chamfers [11]

With the bend test, weld seam quality can be evaluated by either the bend angle or the ram displacement (unfortunately there is no correlation between bend angle and ram displacement). Thus, prior to the test, the operator must determine whether to measure the ram displacement or the bend angle.

Table 15.12 Ram Speed During Bend Tests for Selected Thermoplastics [11]

Material	Test speed [mm/min]
PE-HD	50
PP-R	50
PP-H,-B	20
PVDF	20
PVC-U	10

Figure 15.9 shows the minimum bend angle technique. The bend angle is defined as the difference between the angle before testing and the final angle when the sample fails. The specimen usually fails by a sudden break or by the initiation of a small crack, followed by rapid crack growth. If a specimen can be fully bent (bend angle 160°) without a fracture, it is referred to as a "no failure" sample.

If ram displacement is measured, the displacement at break or at the first sign of visible cracks is recorded. Table 15.13 shows the ram displacement that should be used for various sample thicknesses ("no failure") for a full bend test. The crack initiation can be observed with the aid of a mirror in the critical area.

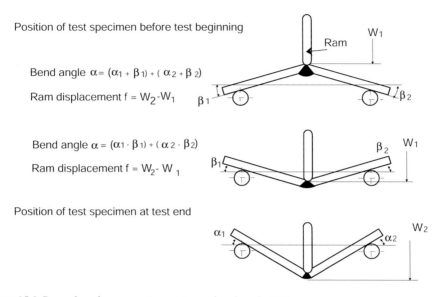

Position of test specimen before test beginning

Bend angle $\alpha = (\alpha_1 + \beta_1) + (\alpha_2 + \beta_2)$

Ram displacement $f = W_2 - W_1$

Bend angle $\alpha = (\alpha_1 - \beta_1) + (\alpha_2 - \beta_2)$

Ram displacement $f = W_2 - W_1$

Position of test specimen at test end

Figure 15.9 *Procedure for measuring minimum bend angle [11]*

Table 15.13 *Ram Displacements Corresponding to a Bend Angle of 160° ("No failure") [11]*

Thickness of Test Specimen s [mm]	Bend Angle [°]	Ram Displacement f [mm]
$3 < s \leq 5$		60
$5 < s \leq 15$		70
$16 < s \leq 20$	160	85
$21 < s \leq 25$		170
$26 < s \leq 30$		150

Whether weld quality is "acceptable" or "unacceptable" is application-dependent; it can be determined by the minimum required bend angle or the minimum required ram displacement. Figure 15.10 shows the minimum bend angle that a "good weld" should be able to withstand as a function of sample thickness for hot gas and extrusion welding [8]. Additional minimum bend angles and ram displacements of other materials are listed in DVS 2203-1. If at least two specimens do not reach the required angle or displacement, two additional test specimens must be tested [11].

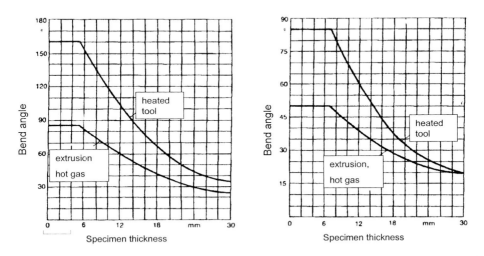

Figure 15.10 Minimum bend angle as a function of specimen thickness [8]

In the field, such as on construction sites or in workshops, screening tests can be performed using the bend test by hand (see Fig. 15.11). In this test, the specimen is bent slowly until fracture or until the specimen is fully bent back onto itself. The bend test by hand is usually limited to a specimen thickness up to 10 mm, because of the manual forces that are required.

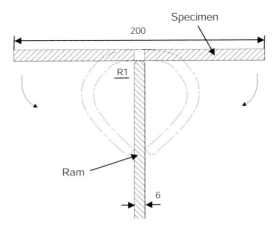

Figure 15.11 Bend test by hand for a first, quick evaluation

It is possible to perform the bend test at different conditions in order to get a better "resolution" or sensitivity of the weld quality characteristics. For example, the so-called "modified" bend test is performed at a lower temperature (e.g., −20 °C). It is also possible to perform the test at a higher test speed. There is evidence of a good correlation between the results of the "modified" bend test and the tensile creep test; in this case the bend angle at the maximal force should be used for comparison instead of bend angle at break [12, 13].

15.2.4   Tensile Creep Test

Long-term weld performance cannot be predicted with short-term tests. This is especially true with the short-term tests such as the tensile test or the bend test. Even with increased testing speed and lowered testing temperature, these tests do not provide information on a weld's long-term performance. The reason for this can be found in the failure mechanisms resulting from short- and long-term loads [14]. However, it is possible to gain some insight into long-term performance with short-term tests. For example, if tensile strength results do not produce sufficient strength to meet the application's load requirements, it can be expected that long-term performance will also not meet the requirements. In contrast, if short-term tensile tests show that the weld quality is sufficient to meet the application's requirements, long-term tests must be completed to determine if the weld will meet the application's requirements over the duration of the of the application's expected service life. One possible long-term test is the tensile creep test, which is detailed in this section.

Because many high-density polyethylene (HDPE) pipes are buried in the ground and inspection is not possible, it is important to predict their weld performance over a long period. While time- and temperature-dependent strength behavior has been determined for HDPE by conducting creep rupture test under internal pressure, these tests do not provide information on long-term weld performance. In a creep rupture test, thermoplastics are loaded with constant internal pressure. The stresses (longitudinal stress), which interact vertically on the weld seam as a result of internal pressure (e.g. in connected pipes), account for about 50% of the circumferential stress (Fig. 15.12). Studies have shown [15]

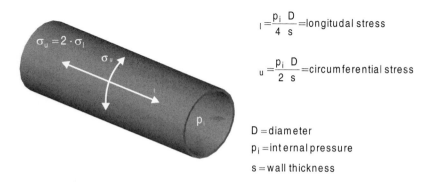

$$\sigma_l = \frac{p_i\, D}{4\, s} = \text{longitudal stress}$$

$$\sigma_u = \frac{p_i\, D}{2\, s} = \text{circumferential stress}$$

$D =$ diameter
$p_i =$ internal pressure
$s =$ wall thickness

Figure 15.12 *State of stress in a pipe*

that the creep rupture behavior of butt-welded pipes is identical to the behavior of non-welded pipes. Thus, long-term properties of welded joints can be determined by creep tensile testing on specimen taken from the un-welded pipe sample axially [16].

The long-term strength behavior of thermoplastics can be defined by the material's creep rupture behavior. Usually, this behavior is determined by a creep rupture test under internal pressure. Fig. 15.13 shows the reference stress logarithm as a function of lifetime logarithm for this type of testing. The greater of the two stresses is defined as reference stress.

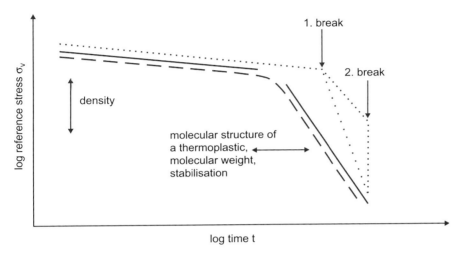

Figure 15.13 *Characteristic creep rupture curve [17]*

The creep rupture curve usually consists of two or three phases with different gradients. In the first phase, the "flat phase", there is very high stress over a short period. In this phase, the specimen fails with a relatively high amount of deformation/strain [18]. If failure occurs in this phase, the failure mechanism is considered "creeping" [19].

The second phase shows a higher gradient or slope. The resulting stresses here are too low to stretch the entire cross-section of a specimen simultaneously. However, molecular chains within the micro-strained-area are quickly pulled from each other. The resulting voids between the molecular chains cause a brittle fracture [19, 20, 21].

A third failure mechanism can be observed in the third phase of the curve. Here, the specimen shows very brittle fractures after longer periods of time. The failure is caused by thermo-oxidative ageing. This failure mechanism is independent of load or stress, but dependent on time and temperature. Thus, this third phase in the creep diagram usually shows a vertical straight line or a very high slope. The time to failure may vary depending on the nature and the quality of stabilizers compounded with the plastic [17].

The creep rupture behavior of thermoplastics can be tested under different environmental test media, usually referred to as the "medium tensile creep test". Different media are selected based on the parts and application requirements. The creep rupture strength can be greatly affected by the medium, which can cause chemical and/or physical failure mechanisms. For instance, oxidizing media, such as nitric acid or sulphuric acid, cause a degradation reaction on polyolefin materials (a break of C-C covalence bonds) [18].

The creep rupture strength of new materials causes very long creep rupture times. Therefore, it has become important to develop a test procedure that allows measuring these properties more easily and quickly.

Increasing the stresses on the sample during testing, usually promotes the specimen to fail quicker with a relatively high amount of deformation (creeping). However, the relatively high amount of deformation is usually considered unacceptable [22]. It is considered a better approach to conduct the tensile creep test in a medium and/or with higher temperatures that can accelerate the failure mechanism. Such a medium must not be chemically aggressive, and must not have a swelling effect on the material. Surfactants are often used to accelerate the creep rupture test since they do not exhibit these properties and have a lower surface tension compared to water [23]. Table 15.14 shows a number of surfactants and the individual time factors, respectively. The time factor f_t is the approximated shortening effect of a surfactant compared to that of water.

Creep tensile tests can be used to predict the long-term quality of a welded joint, similar to creep burst tests. Because the stress/strain curves of the base material and the welded material are usually very similar, it can be implied that the failure mechanisms are also similar. The long-term quality of a welded joint is usually specified by the long-term welding factor f_s. The long term welding factor can be determined by the relationship of the two stress values, which relate the time to failure for the welded joint and the reference material (Fig. 15.14) [16].

The creep tensile strength of a tested tensile bar is usually seven to ten times higher than the observed strength of comparable pipes under internal pressure (creep rupture test) [22]. Therefore, it is necessary to conduct the creep tensile test with a surfactant at elevated temperatures in order to accelerate the failure. Usually, the surfactants with time-shortening properties at high temperatures will produce welding factors similar to those observed with water.

$$f_t = \frac{\text{time until failure in medium}}{\text{time until failure in water}}$$

Table 15.14   *Time Factors of Selected Surfactants [22]*

Medium	Concentration	Time Factor f_t
Anionic surfactants	0.0002 %	2
Anionic surfactants	0.043 %	5
Arkopal N 100	2 %	25 – 30
Dodigen 179/180	1 %	250

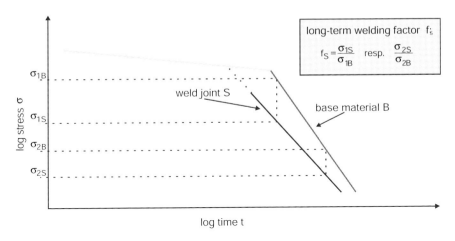

Figure 15.14 *Evaluation of the long-term welding factor*

Testing Procedure:

Ideally, the stresses induced during a long-term test of a welded joint are equivalent to the actual stresses induced when the part is in service. A welded specimen should be tested (with or without welding bead) by exposing it to suitable temperatures and environmental conditions. The test should not result in over-stressing/straining nor should it promote ductile fracture [22]. In order to provide data for a statistical evaluation, six welded and six base material specimens should be tested under the same conditions. The average value will be calculated as the geometrical average value, based on the logarithmic creep rupture time.

The specimens used for creep tensile testing vary in geometry and size depending on the welding technique as well as on the application requirements. Usually, welded plates and pipes are tested with samples cut from the application to form a rectangle- or shoulder specimen (the shoulder is to prevent failure occurring in the clamping). It is important to record the location from which the sample was taken, as well as the technique used to cut the sample. The cutting/machining technique can influence the sample's surface (ridges, notches, or thermal damage), which may affect the test results. For example, specimens cut or planed by a water jet reach higher strengths [22, 24] than specimens prepared by mechanical machining.

Figure 15.15 shows a set-up for a creep tensile test. In this case, the welded specimens are affixed with a clamp in a basin containing the testing medium. The welded specimens are loaded with testing weights. The exact loading force applied to the samples can be determined by using a load cell. Additionally, a circulation pump is often used to mix the testing medium.

During the test, the weld seam must be placed equally distanced from each clamp. After installation, the specimens should remain in the testing medium for 60 to 120 minutes in order to assure temperature equilibrium throughout the sample and medium. The length of time to reach equilibrium depends on the sample thickness. While the sample is immersed in the medium, the specimens must not be mechanically loaded (see also [22]), and impact loading of the samples should be avoided, instead the load should be applied gradually.

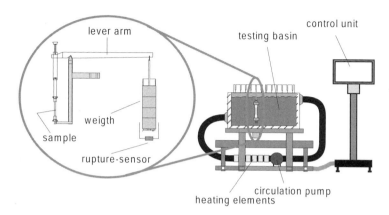

Figure 15.15 *Test set-up for creep tensile test*

Creep tensile tests should be conducted in a medium at a constant temperature under a constant stress. The temperature of the testing medium must be strictly controlled during the test, because a deviation in medium temperature of 1°C at a testing temperature of 80 °C can extend or shorten the creep rupture time by 13 % [25].

Usually the resulting creep rupture strength data is plotted as a function of time. This allows the projection or interpolation of long-term welding seam quality (see Fig. 15.13). The creep rupture strength curve is determined at various stresses.

When evaluating the long-term welding factor, only the second phase of the creep rupture strength provides important information (Fig. 15.13). In addition, visual inspection of the fracture surface can be used to gain insight on the performance of the weld. For example, the amount of brittle fracture surface should be less than 30% of the entire cross-section [22].

In general, most base materials show defects of some kind, e.g., non-homogeneity, weak points or cracks produced from molding, mixing, or extrusion. These defects can slowly grow and promote failure when exposed to stress. The following issues can also promote crack growth:

- Ridges on the surface of a component, caused by inappropriate processing/manufacturing,
- Rocks, which press on an inappropriately installed pipe,
- Local overload, induced by notches, due to bad part design
- Notches at the edges of the weld bead (Fig. 15.16) [19],
- Structural characteristics of a weld bead [26].

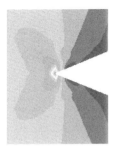

Figure 15.16 *Stress concentration at edge of weld bead*

Crack growth can be initiated not only by loading of a weld at a notch or a defect location, but also by residual stresses within the weld [27]. The degree to which a defect or crack will promote failure depends on its size, notch radius, and the material's resistance against slow crack growth. The long-term behavior of a material as well as that of a weld seam made from this material is strongly dependent on its resistance against slow crack growth. Slow crack growth can be determined by the so-called Full Notch Creep Test (FNCT).

The set-up for an FNCT is similar to that for a creep tensile test. Figure 15.17 shows cross-sections of typical test specimens. The four sides of the specimen should each be notched in the middle to similar depths.

The results of the creep rupture test under internal pressure and the tensile creep test differ considerably [25]. The large artificial crack introduced with the FNCT can result in small deviations of the tests results. Because of the large artificial crack, the creep rupture behavior is not influenced by the crack initiation behavior of the surface, but only by the crack propagation behavior. The FNCT is a time-dependent testing method, in which the specimens usually fail by mechanisms similar to those during creep rupture testing. The FNCT is faster than other long-term tests, because the slow crack initiation step is artificially introduced, reducing the testing time by a factor of 30 [27].

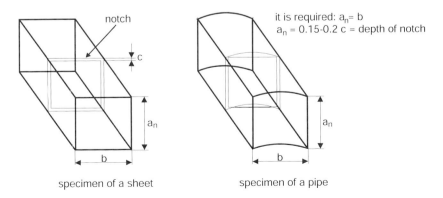

Figure 15.17 *Notched specimens for the FNCT*

It is important to point out again that the FNCT characterizes the material without considering the crack initiation, whereas the creep rupture test measures the real creep rupture behavior of a component based on the entire sample [25]. Usually, the creep rupture test is the preferred procedure to determine long-term weld quality, which is characterized by the long-term welding factor. The fact that a base material has good long-term properties does not mean automatically that weld seams made from this material also show good long-term properties. The material may have a small resistance against slow crack growth, but due to, e.g., the notch effect of the weld bead, early failure may occur. Thus, in summary, the FNCT measures, with comparably less effort, whether a material is suitable for the application of a durable welded joint. It also helps to determine whether the weld will fail, due to the material's low resistance against slow crack growth [28].

15.3   References

1. N.N. DVS 2202-1 –Imperfections in Thermoplastic Welding Joints; Features, Descriptions, Evaluation (1999) DVS-Verlag GmbH, Düsseldorf/Germany

2. Uebbing, M., Fügen von Kunststoffen – Leitfaden für Fertigung und Konstruktion (1998) DVS-Verlag, Düsseldorf/Germany

3. N.N. DVS 2206 – Prüfen von Bauteilen und Konstruktionen aus thermoplastischen Kunststoffen (1995) DVS-Verlag GmbH, Düsseldorf/Germany

4. N.N. AWS G1.2M, Specification for Standardized Ultrasonic Welding Test Specimen for Thermoplastics (1999) ANSI,

5. Grimm, R., Yeh, H., Infrared Welding of Thermoplastics, Colored Pigment and Carbon Black Levels on Transmission of Infrared Radiation, ANTEC '98 Conference Proceedings, Society of Plastics Engineers, Brookfield/CT

6. N.N. DVS 2203-5 – Testing of Welded Joints of Thermoplastic Materials: Tensile Test, (1985) DVS-Verlag GmbH, Düsseldorf/Germany

7. Menges, G., Schmachtenberg, E., Knausenberger, R., Ein schneller Weg zur Ermittlung des Langzeitverhaltens von Schweißnähten (1983) DVS-Berichte 84, Düsseldorf/Germany

8. N.N. DVS 2203-1 – Testing of Welded Joints of Thermoplastic Semi-Finished Products – Test Methods – Requirements (1986) DVS-Verlag GmbH, Düsseldorf/Germany

9. Benatar, A., Poopat, B., Wu, C., Hot Plate Welding of Polypropylene and Talc Reinforced Polypropylene Composites, Conference Proceedings, ANTEC 1999, pp. 1381-85, Society of Plastic Engineers, Brookfield, CT., USA

10. N.N. DVS 2203-3 – Prüfen von Schweißverbindungen aus thermoplastischen Kunst-stoffen: Schlagzugversuch. Testing of Welded Joints of Thermoplastic Plates And Tubes (1985) DVS-Verlag GmbH, Düsseldorf/Germany,

11. N.N. DVS 2203-5 – Testing of Welded Joints of Thermoplastic Plates And Tubes: Technological Bend Test (1999) DVS-Verlag GmbH, Düsseldorf/Germany

12. Tappe, P., Modellgesetze beim Heizelement-Stumpfschweißen teilkristalliner Thermo-plaste, Ph.d.-thesis (1989) University of Paderborn/Germany

13. Hessel, J., John, P., Heizelementschweißen von HDPE-Rohren und -Platten, *Kunst-stoffe*, **75**, 1985, 10

14. Gehde, M., Langzeit- und Versagensverhalten von Extrusionsschweißverbindungen, *Schweißen und Schneiden*, **48**, 1996, pp. 278/285

15. Diedrich, G., Gaube, E., Zeitstandfestigkeit und Langzeit-Schweißfaktoren von geschweißten Rohren und Platten aus Hart-Polyäthylen und Polypropylen, *Kunststoffe*, **63**, 1973, pp. 793/797

16. N.N. DVS 2203-4 – Testing of Welded Joints of Thermoplastics Materials: Tensile Creep Test (1997) DVS-Verlag GmbH, Düsseldorf/Germany

17. Gebler, H., Langzeitverhalten und Alter von PE-HD-Rohren, *Kunststoffe*, **79**, 1989, pp. 823/826

18. Diedrich, G., Kempe, B., Graf, K., Zeitstandfestigkeit von Rohren aus Polyethylen (HDPE) und Polypropylen (PP) unter Chemikalieneinwirkung, *Kunststoffe*, **69**, 1979, pp. 470/476

19. Lecht, R., Schulte, U., Welches Material? Rohr- und Behälterwerkstoffe mit dem Zeitstand-Zugversuch (FNCT) auswählen, *Materialprüfung*, **40**, 1998, pp. 399/402

20. Gaube, E., Kausch, H., Bruchtheorien bei der industriellen Anwendung von Thermo-plasten und glasfaserverstärkten Kunststoffen, *Kunststoffe*, **63**, 1973, pp. 391/397

21. Gaube, E., Müller, W., Einfluß des Werkstoffs und der Verarbeitung auf die Zeitstand-festigkeit von Rohren aus HDPE, *Kunststoffe*, **72**, 1982, pp.297/299

22. Hessel, J., Mauer, E., Zeitstandzugprüfung in wässriger Netzmittellösung. *Material-prüfung*, **36**, 1994, pp. 240/243

23. Hessel, J., Zeitstandverhalten von Polyethylen unter dem Einfluß lokal konzentrierter Spannungen. *3R International*, **34**, 1995, pp. 573/579

24. Hessel, J. et al., Neues Konzept zur Lebensdauerbestimmung an Kunststoff-Mantel-rohr-Schweißverbindungen aus Polyethylen, *3R International*, **34**, 1995, pp. 94/101

25. Fleißner, M., Langsames Risswachstum und Zeitstandfestigkeit von Rohren aus Poly-ethylen, *Kunststoffe*, **77**, 1987, pp. 45/50

26. Gehde, M., Analysis of The Deformation of Polypropylene Hot-Tool Butt Welds, *Polymer Engineering and Science*, **32**, 1992, pp. 586/592

27. Lecht, R. et al., Neues Verfahren zur Prüfung des Langzeitverhaltens und Schweißbarkeit von Kunststoffen für Rohre, *3R International,* **34**, 1995, pp. 612/615

28. Lecht, R., Schulte, U., Was nützt der FNCT dem Praktiker? *Kunststoffe,* **87**, 1997, pp. 1405/1406

All DVS-guidelines for testing plastic welds are available on CD-Rom in German and English from DVS Verlag GmbH, Aachener Straße 172, 40233 Düsseldorf/Germany, "http:\\www.dvs-verlag.de", e-mail: "verlag@dvs-hg.de"

Index